I0484382

United States
Environmental Protection
Agency

Climate Change Indicators in the United States, 2012

2ND EDITION

Find Us Online

EPA's updated climate change website now features a more user-friendly interface and downloadable images and figures. To view the latest information about EPA's climate change indicators, along with the corresponding technical documentation, or to suggest new indicators for future reports, please visit EPA's website at: www.epa.gov/climatechange/indicators or send a message to: climateindicators@epa.gov.

Contents

Acknowledgments

This report was developed by EPA's Office of Atmospheric Programs, Climate Change Division, with support from the Office Research and Development and the Office of Water. It also reflects the contributions and collaboration of many other individuals. EPA received essential support from scientists and communications experts at a number of federal agencies, universities, nongovernmental organizations, and international institutions.

Data Contributors and Indicator Reviewers

U.S. Federal Agencies

Centers for Disease Control and Prevention	George Luber
National Aeronautics and Space Administration	Joey Comiso
National Oceanic and Atmospheric Administration	*Climate Prediction Center:* Gerry Bell
	Earth System Research Laboratory: Steve Montzka
	National Climatic Data Center: Deke Arndt, Karin Gleason, Boyin Huang
	National Oceanographic Data Center: Sydney Levitus
	National Ocean Service: Chris Zervas
	Office of Oceanic and Atmospheric Research: Libby Jewett
	Pacific Marine Environmental Laboratory: Richard Feely, Chris Sabine
National Snow and Ice Data Center	Walt Meier
U.S. Department of Agriculture	Lewis Ziska
U.S. Geological Survey	*Alaska Science Center:* Shad O'Neel
	Maine Water Science Center: Robert Dudley, Glenn Hodgkins
	New York Water Science Center: Mike McHale
	Washington Water Science Center: Bill Bidlake, Mark Savoca

Universities, Nongovernmental Organizations, and International Institutions

Bermuda Institute of Ocean Sciences	Nick Bates
California Department of Public Health	Paul English
Commonwealth Scientific and Industrial Research Organisation	John Church, Catia Domingues, Neil White
Georgia Institute of Technology	Ray Wang
Japan Agency for Marine-Earth Science and Technology	Masayoshi Ishii
Massachusetts Institute of Technology	Kerry Emanuel
North Carolina State University	Ken Kunkel
Rutgers University Global Snow Lab	David Robinson
University of Nebraska-Lincoln	Song Feng
Universidad de las Palmas de Gran Canaria	Melchor González-Dávila
University of Wisconsin-Madison	Corinna Gries
University of Wisconsin-Milwaukee	Mark Schwartz
USA National Phenology Network	Jake Weltzin
Woods Hole Oceanographic Institution	Sarah Cooley
World Glacier Monitoring Service	Michael Zemp
World Resources Institute	Tom Damassa

Peer Review

The report included an external peer review consisting of 12 expert reviewers: Michael C. MacCracken, Tanja Srebotnjak, Dan Tunstall, Paul Kirshen, Thomas R. Knutson, Gerald Meehl, Steven Nerem, W. Tad Pfeffer, Michael J. Prather, David Schimel, Joel D. Schwartz, and Claudia Tebaldi.

Report Production and Design

Support for the report's production and design was provided by Eastern Research Group, Inc. (ERG).

Introduction

The Earth's climate is changing. Scientists are confident that many of the observed changes in the climate can be linked to the increase in greenhouse gases in the atmosphere, caused largely by people burning fossil fuels to generate electricity, heat and cool buildings, and power vehicles (see "The Greenhouse Effect" below to learn about how these gases trap heat). Current and future emissions will continue to increase the levels of these gases in our atmosphere for the foreseeable future.

One way to track and communicate the causes and effects of climate change is through the use of indicators. An indicator, such as a record of Arctic sea ice extent, represents the state or trend of certain environmental conditions over a given area and a specified period of time. Scientists, analysts, decision-makers, and others use environmental indicators, including those related to climate, to help monitor environmental trends over time, track key factors that influence the environment, and identify effects on ecosystems and society.

The **climate change indicators** in this report present compelling evidence that the composition of the atmosphere and many fundamental measures of climate in the United States are changing. Temperatures are rising, snow and rainfall patterns are shifting, and more extreme climate events—like heavy rainstorms and record high temperatures—are taking place. Similar changes are occurring around the world.

These observed changes affect people and the environment in important ways. For example, sea levels are rising, glaciers are melting, and plant and animal life cycles are changing. These types of changes can bring about fundamental disruptions in ecosystems, affecting plant and animal populations, communities, and biodiversity. Such changes can also affect society, including where people can live, what kinds of crops farmers can grow, and what kinds of businesses can thrive in certain areas.

Indicators of climate change are expected to become even more numerous and depict even clearer trends in the future. Looking ahead, the U.S. Environmental Protection Agency (EPA) will continue to work in partnership with other agencies, organizations, and individuals to collect and communicate useful data and to inform policies and programs based on this knowledge.

What Is Climate Change?

Climate change refers to any significant change in measures of climate (such as temperature or precipitation) lasting for an extended period (decades or longer). Climate change may result from natural factors and processes and from human activities.

Global warming is a term often used interchangeably with the term "climate change," but they are not the same thing. Global warming refers to an average increase in the temperature of the atmosphere near the Earth's surface. Global warming is just one aspect of global climate change, albeit a very important one.

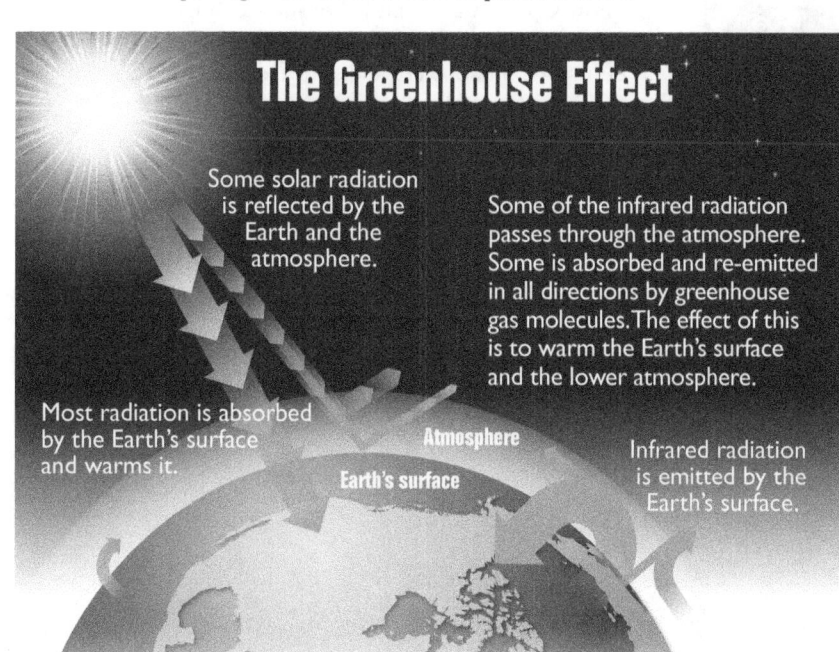

The Greenhouse Effect

Some solar radiation is reflected by the Earth and the atmosphere.

Some of the infrared radiation passes through the atmosphere. Some is absorbed and re-emitted in all directions by greenhouse gas molecules. The effect of this is to warm the Earth's surface and the lower atmosphere.

Most radiation is absorbed by the Earth's surface and warms it.

Atmosphere

Earth's surface

Infrared radiation is emitted by the Earth's surface.

About This Report

Climate Change Indicators in the United States, 2012, presents 26 indicators to help readers better understand observed trends related to the causes and effects of climate change. This document updates a report published by EPA in 2010.

Various government agencies, academic institutions, and other organizations contributed data critical to the development of this report. EPA also received feedback from a diverse group of scientists, researchers, and communications experts in the public and private sectors. This feedback helped to inform the content and new features of this 2012 report. All of the indicators in this report are based on data that have been collected and compiled according to protocols accepted by the scientific community. The indicators were chosen using a standard set of criteria that considered usefulness, objectivity, data quality, transparency, ability to meaningfully communicate, and relevance to climate change. In addition, the report was peer-reviewed by independent technical experts.

EPA's Greenhouse Gas Reporting Program

EPA is now collecting facility-level data on U.S. greenhouse gas emissions and other relevant information under the Greenhouse Gas Reporting Program. This program requires annual reporting of greenhouse gas data from large emissions sources across a range of industry sectors, as well as suppliers of products that would emit greenhouse gases if released or combusted. This new information will help inform the annual *Inventory of U.S. Greenhouse Gas Emissions and Sinks,* which currently serves as the data source for the U.S. Greenhouse Gas Emissions indicator.

For more information, see: www.epa.gov/climatechange/emissions/ghgdata.

Who Is This Report For?

Climate Change Indicators in the United States, 2012, is written with the primary goal of informing readers' understanding of climate change. In addition to presenting climate change observations and trends in the United States and globally, this report highlights the far-reaching significance of these changes and their possible consequences for people, the environment, and society.

This report is also designed to be useful for scientists, analysts, decision-makers, educators, and others who can use climate change indicators as a tool for:

- Assessing trends in environmental quality, factors that influence the environment, and effects on ecosystems and society.

- Effectively supporting science-based decision-making and communication.

- Evaluating existing and future climate-related policies and programs.

What's New?

The 2012 report reflects the following new features and changes:

- *Three new indicators:* **Snowfall, Streamflow,** and **Ragweed Pollen Season**. These additions provide further evidence of climate change and its effects that are being felt by different kinds of ecosystems, as well as by society.

- *Expanded indicators:* **Arctic Sea Ice** was expanded to show changes in the age of ice and **Snow Cover** was expanded to show changes in snow cover for particular seasons. Several decades of historical data were added to **Drought**, and the 2010 **Heat Waves** indicator was converted to **High and Low Temperatures**.

- *Updated indicators:* Nearly all indicators have been updated with additional years of data that have become available since the last report.

- *Regional perspectives:* Several indicators include maps that show how trends vary by region.

A Roadmap to the Report

Most of the indicators in this report focus on the United States, but some include global trends to provide context or a basis for comparison, while others have a regional focus. Geographic coverage depends on data availability and the nature of what is being measured. For example, greenhouse gas concentrations in the atmosphere are studied on a global scale. The indicators span a range of time periods, depending on data availability. Each indicator features five elements:

- One or more graphics depicting changes over time. Some indicators consist of a single metric, while others present multiple metrics (for example, the Drought indicator shows two different ways of calculating drought).
- Key points about what the graphics show.
- Background on how the indicator relates to climate change.
- Information about how the indicator was developed.
- Factors that influence the potential to draw valid conclusions from the indicator.

The indicators are divided into five chapters:

Greenhouse Gases: Greenhouse gases from human activities are responsible for the largest share of climate change since the mid-20th century. The indicators in this chapter characterize emissions of the major greenhouse gases resulting from human activities, the concentrations of these gases in the atmosphere, and how emissions and concentrations have changed over time.

Weather and Climate: Rising global average temperature is linked to certain widespread changes in weather patterns, which in turn lead to changes in the Earth's climate (the average weather over time). This chapter focuses on indicators related to weather and climate, including temperature, precipitation, storms, and droughts.

Oceans: The world's oceans have a two-way relationship with weather and climate. The oceans influence the weather on local to global scales, while changes in climate can fundamentally alter certain properties of the ocean. This chapter examines trends in ocean characteristics that relate to climate change, such as heat storage, temperature, and sea level.

Snow and Ice: Climate change can alter the Earth's snow- and ice-covered areas. These changes, in turn, can affect air temperatures, sea levels, ocean currents, and storm patterns. This chapter focuses on trends in glaciers and sea ice, snowfall, extent and depth of snow cover, and the freezing and thawing of oceans and lakes.

Society and Ecosystems: Changes in the Earth's climate can affect public health, agriculture, water supplies, energy production and use, land use and development, and recreation. Climate change can also disrupt the functioning of ecosystems and increase the risk of harm or even extinction for some species. This chapter looks at some of the ways that climate change is affecting society and ecosystems, including changes in allergy seasons, heat-related deaths, streamflows, and bird migration patterns.

The report concludes with a discussion on **climate change indicators and health**, as further development of human health indicators is of increasing importance. Climate change impacts associated with human health include expected increases in heat-related illness and death, worsening air quality, and likely increases in the frequency and strength of certain extreme events such as floods, droughts, and storms. Climate change may also allow some diseases to spread more easily. People most vulnerable to health impacts include the poor, the elderly, those already in poor health, the disabled, and indigenous populations. EPA plans to explore opportunities to work with climate and health experts to develop indicators that communicate the effects of climate change on health and society more broadly.

EPA has compiled an accompanying **technical support document** containing more detailed information about each indicator, including data sources, data collection methods, calculations, statistical considerations, and sources of uncertainty. This document also describes EPA's approach and criteria for selecting indicators for the report. This information is available on EPA's website at: www.epa.gov/climatechange/indicators.

Additional resources that can provide readers with more information appear at the end of the report (see Climate Change Resources on p. 75).

Looking Ahead

As new and more comprehensive indicator data become available, EPA plans to continue to periodically update the indicators presented in this report to document climate change and its effects.

Summary of Key Points

Greenhouse Gases

Human activities have substantially increased the amount of greenhouse gases in the atmosphere, leading to warming of the climate and many other changes around the world—effects that will persist over a long time.

U.S. Greenhouse Gas Emissions. In the United States, greenhouse gas emissions caused by human activities increased by 10 percent from 1990 to 2010. Carbon dioxide accounts for most of the nation's emissions and most of this increase. Electricity generation is the largest source of greenhouse gas emissions in the United States, followed by transportation. Emissions per person have decreased slightly in the last few years.

Global Greenhouse Gas Emissions. Worldwide, emissions of greenhouse gases from human activities increased by 26 percent from 1990 to 2005. Emissions of carbon dioxide, which account for nearly three-fourths of total emissions, increased by 31 percent over this period. As with the United States, the majority of the world's emissions result from energy production and use.

Atmospheric Concentrations of Greenhouse Gases. Concentrations of carbon dioxide and other greenhouse gases in the atmosphere have increased since the beginning of the industrial era. Almost all of this increase is attributable to human activities.[1] Historical measurements show that current levels of many greenhouse gases are higher than any levels recorded for hundreds of thousands of years, even after accounting for natural fluctuations.

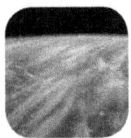

Climate Forcing. Climate or "radiative" forcing is the measurement of how substances such as greenhouse gases affect the amount of energy absorbed by the atmosphere. An increase in radiative forcing means a heating effect, which leads to warming, while a decrease in forcing produces cooling. From 1990 to 2011, the total radiative forcing from greenhouse gases added by humans to the Earth's atmosphere increased by 30 percent. Carbon dioxide has accounted for approximately 80 percent of this increase.

Weather and Climate

Variations in weather and climate cause changes in temperature, precipitation, and extreme event patterns, which can directly or indirectly affect many aspects of society.

 **U.S. and Global Temperature.** Average temperatures have risen across the contiguous 48 states since 1901, with an increased rate of warming over the past 30 years. Seven of the top 10 warmest years on record have occurred since 1990. Recent compilations of the change in global average temperatures show a similar trend, and 2001–2010 was the warmest decade on record worldwide. Within the United States, temperatures in parts of the North, the West, and Alaska have increased the most.

 **High and Low Temperatures.** Since the 1970s, unusually hot summer temperatures have become more common in the United States, and heat waves have become more frequent. In contrast, extremely cold winter temperatures have become less common. The most recent decade has had twice as many record high temperatures as record lows. The most severe heat waves in U.S. history remain those that occurred during the "Dust Bowl" in the 1930s.

 **U.S. and Global Precipitation.** Total annual precipitation has increased in the United States and over land areas worldwide. Since 1901, precipitation has increased at an average rate of nearly 6 percent per century in the contiguous 48 states and more than 2 percent per century over land areas worldwide. However, shifting weather patterns have caused certain areas, such as Hawaii and parts of the Southwest, to experience less precipitation than usual.

 **Heavy Precipitation.** In recent years, a higher percentage of precipitation in the United States has come in the form of intense single-day events. Nationwide, eight of the top 10 years for extreme one-day precipitation events have occurred since 1990. The occurrence of abnormally high annual precipitation totals (as defined by the National Oceanic and Atmospheric Administration) has also increased.

 **Drought.** Average drought conditions across the nation have varied since records began in 1895. The 1930s and 1950s saw the most widespread droughts, while the last 50 years have generally been wetter than average. However, specific trends vary by region. A more detailed index developed recently shows that between 2000 and 2011, roughly 30 to 60 percent of the U.S. land area experienced drought conditions at any given time.

 **Tropical Cyclone Activity.** Tropical storm activity in the Atlantic Ocean, Caribbean, and Gulf of Mexico has increased during the past 20 years. This increase is closely related to variations in sea surface temperature in the tropical Atlantic. However, changes in observation methods over time make it difficult to know for sure whether a long-term increase has occurred. Records collected since the late 1800s suggest that the actual number of hurricanes per year has not increased.

Oceans

Changes in ocean temperature, sea level, and seawater chemistry have implications for coastal communities and could substantially alter the biodiversity and productivity of ocean ecosystems.

Ocean Heat. Several studies have shown that the amount of heat stored in the ocean has increased substantially since the 1950s. Ocean heat content not only determines sea surface temperature, but also affects sea level and currents.

Sea Surface Temperature. Ocean surface temperatures increased around the world over the 20th century. Even with some year-to-year variation, the overall increase is statistically significant, and sea surface temperatures have been higher during the past three decades than at any other time since reliable observations began in the late 1800s.

Sea Level. When averaged over all the world's oceans, sea level has increased at a rate of roughly seven-tenths of an inch per decade since 1880. The rate of increase has accelerated in recent years to more than an inch per decade. Changes in sea level relative to the height of the land vary widely because the land itself moves. Along the U.S. coastline, sea level has risen the most relative to the land along the Mid-Atlantic coast and parts of the Gulf Coast. Sea level has decreased relative to the land in parts of Alaska and the Northwest.

Ocean Acidity. The ocean has become more acidic over the past 20 years because of increased levels of atmospheric carbon dioxide, which in turn dissolves in the water. Higher acidity has led to decreased availability of minerals such as aragonite, which is an important form of calcium carbonate that many marine animals use to build their skeletons and shells.

Snow and Ice

Climate change can dramatically alter the Earth's snow- and ice-covered areas, affecting vegetation and wildlife, water supplies and transportation, and communities in Arctic regions.

Arctic Sea Ice. Part of the Arctic Ocean is covered by ice year-round. The area covered by ice is typically smallest in September, after the summer melting season. The minimum extent of Arctic sea ice has decreased over time, and in September 2012 it was the smallest on record. Arctic ice has also become thinner, which makes it more vulnerable to additional melting.

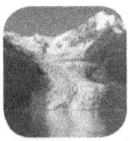

Glaciers. Glaciers in the United States and around the world have generally shrunk since the 1960s, and the rate at which glaciers are melting has accelerated over the last decade. The loss of ice from glaciers has contributed to the observed rise in sea level.

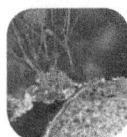

Lake Ice. Lakes in the northern United States generally appear to be freezing later and thawing earlier than they did in the 1800s and early 1900s. The length of time that lakes stay frozen has decreased at an average rate of one to two days per decade.

Snowfall. Total snowfall has decreased in most parts of the country since widespread records began in 1930. One reason for this decline is that more than three-fourths of the locations studied have seen more winter precipitation fall in the form of rain instead of snow.

 **Snow Cover.** The portion of North America covered by snow has decreased somewhat since 1972, based on weekly measurements taken throughout the year. However, there has been much year-to-year variability. During the years 2002–2011, the average area covered by snow was 3 percent (roughly 100,000 square miles) smaller than the average extent of snow cover during the first 10 years of measurement (1972–1981).

 **Snowpack.** The depth of snow on the ground (snowpack) in early spring decreased at most measurement sites—some by more than 75 percent—between 1950 and 2000. However, a few locations in the western United States and Canada saw an increase in spring snowpack.

Society and Ecosystems

Climate change could require adaptation on larger and faster scales than in the past, presenting challenges to human well-being, the economy, and natural ecosystems.

 **Streamflow.** Changes in temperature, precipitation, snowpack, and glaciers can affect the amount of water carried by rivers and streams and the timing of peak flow. Over the last 70 years, minimum and maximum flows have changed in many parts of the country—some higher, some lower. Three-fifths of the rivers and streams measured show peak winter-spring runoff happening at least five days earlier than it did in the past.

 **Ragweed Pollen Season.** Warmer temperatures and later fall frosts allow ragweed plants to produce pollen later into the year, potentially prolonging allergy season for millions of people. The length of ragweed pollen season has increased at eight out of 10 locations studied in the central United States and Canada since 1995. The change becomes more pronounced from south to north.

 **Length of Growing Season.** The average length of the growing season in the contiguous 48 states has increased by nearly two weeks since the beginning of the 20th century. A particularly large and steady increase has occurred over the last 30 years. The observed changes reflect earlier spring warming as well as later arrival of fall frosts. The length of the growing season has increased more rapidly in the West than in the East.

 **Leaf and Bloom Dates.** Leaf growth and flower blooms are examples of natural events whose timing can be influenced by climate change. Observations of lilacs and honeysuckles in the contiguous 48 states suggest that first leaf growth is now occurring a few days earlier than it did in the early 1900s. Lilac and honeysuckle bloom dates vary greatly from year to year, which makes it difficult to determine whether a statistically meaningful change has taken place.

 **Bird Wintering Ranges.** Some birds shift their range or alter their migration habits to adapt to changes in temperature or other environmental conditions. Long-term studies have found that bird species in North America have shifted their wintering grounds northward by an average of 35 miles since 1966, with a few species shifting by several hundred miles. On average, bird species have also moved their wintering grounds farther from the coast, consistent with rising inland temperatures.

 **Heat-Related Deaths.** Over the past three decades, more than 7,000 Americans were reported to have died as a direct result of heat-related illnesses, such as heat stroke. The annual death rate rises when accounting for other deaths in which heat was reported as a contributing factor. Considerable year-to-year variability in the data and certain limitations of this indicator make it difficult to determine whether the United States has experienced long-term trends in the number of deaths classified as "heat-related."

Greenhouse

Major Greenhouse Gases Associated with Human Activities

Greenhouse gas	How it's produced	Average lifetime in the atmosphere	100-year global warming potential
Carbon dioxide	Emitted primarily through the burning of fossil fuels (oil, natural gas, and coal), solid waste, and trees and wood products. Changes in land use also play a role. Deforestation and soil degradation add carbon dioxide to the atmosphere, while forest regrowth takes it out of the atmosphere.	see below*	1
Methane	Emitted during the production and transport of coal, natural gas, and oil. Methane emissions also result from livestock and agricultural practices and from the anaerobic decay of organic waste in municipal solid waste landfills.	12 years	21
Nitrous oxide	Emitted during agricultural and industrial activities, as well as during combustion of fossil fuels and solid waste.	114 years	310
Fluorinated gases	A group of gases that includes hydrofluorocarbons, perfluorocarbons, and sulfur hexafluoride, among other chemicals. These gases are emitted from a variety of industrial processes and commercial and household uses, and do not occur naturally. Sometimes used as substitutes for ozone-depleting substances such as chlorofluorocarbons (CFCs).	A few weeks to thousands of years	Varies (the highest is sulfur hexafluoride at 23,900)

This table shows 100-year global warming potentials, which describe the effects that occur over a period of 100 years after a particular mass of a gas is emitted. EPA uses global warming potentials from the Intergovernmental Panel on Climate Change's (IPCC's) Second Assessment Report,[1] as countries have agreed to do under current international guidelines within the United Nations Framework Convention on Climate Change (UNFCCC). Lifetimes come from the IPCC's Fourth Assessment Report.[2]

* Carbon dioxide's lifetime is poorly defined because the gas is not destroyed over time, but instead moves among different parts of the ocean–atmosphere–land system. Some of the excess carbon dioxide will be absorbed quickly (for example, by the ocean surface), but some will remain in the atmosphere for thousands of years, due in part to the very slow process by which carbon is transferred to ocean sediments.

INDICATORS IN THIS CHAPTER

U.S. Greenhouse Gas Emissions

Global Greenhouse Gas Emissions

Atmospheric Concentrations of Greenhouse Gases

Gases

Energy from the sun drives the Earth's weather and climate. The Earth absorbs some of the energy it receives from the sun and radiates the rest back toward space. However, certain gases in the atmosphere, called greenhouse gases, absorb some of the energy radiated from the Earth and trap it in the atmosphere. These gases essentially act as a blanket, making the Earth's surface warmer than it otherwise would be. While this "greenhouse effect" occurs naturally, making life as we know it possible, human activities in the past century have substantially increased the amount of greenhouse gases in the atmosphere, causing the atmosphere to trap more heat and leading to changes in the Earth's climate.

What is happening?

The major greenhouse gases emitted into the atmosphere through human activities are carbon dioxide, methane, nitrous oxide, and fluorinated gases (see Major Greenhouse Gases Associated With Human Activities on p.10). Some of these gases are produced almost entirely by human activities; others come from a combination of natural sources and human activities.

Many of the major greenhouse gases can remain in the atmosphere for tens to hundreds of years after being released. They become globally mixed in the lower atmosphere, reflecting contributions from emissions sources worldwide.

Several factors determine how strongly a particular greenhouse gas will affect the Earth's climate.

One factor is the length of time that the gas remains in the atmosphere. A second factor is each gas's unique ability to absorb energy. By considering both of these factors, scientists calculate a gas's global warming potential, as compared to an equivalent mass of carbon dioxide (which is defined by a global warming potential equal to 1).

Why does it matter?

As greenhouse gas emissions from human activities increase, they contribute to more warming of the climate, leading to many other changes around the world—in the atmosphere, on land, and in the oceans. These changes will have both positive and negative effects on people, plants, and animals. Because many of the major greenhouse gases can stay in the atmosphere for tens to hundreds of years after being released, their warming effects on the climate will persist over a long time.

Gases and Substances Not Included in This Report

This report addresses most of the major, well-mixed greenhouse gases that contribute to warming of the climate. The report does not address trends in emissions or concentrations of substances with shorter atmospheric lifetimes (i.e., less than a year) that are also relevant to climate change, such as ozone in the lower atmosphere, pollutants that lead to ozone formation, water vapor, and aerosols (atmospheric particles) such as black carbon and sulfates. These substances may be considered for future editions of this report.

For detailed information about data used in these indicators, see the online technical documentation at: www.epa.gov/climatechange/indicators.

Climate Forcing

U.S. Greenhouse Gas Emissions

This indicator describes emissions of greenhouse gases in the United States.

Background

A number of factors influence the quantities of greenhouse gases released into the atmosphere, including economic activity, population, consumption patterns, energy prices, land use, and technology. There are several ways to track these emissions. In addition to tracking overall emissions and emissions from specific industrial sectors in absolute terms, many countries also track emissions per capita.

About the Indicator

This indicator focuses on emissions of carbon dioxide, methane, nitrous oxide, and several fluorinated gases—all important greenhouse gases that are influenced by human activities. These particular gases are covered under the United Nations Framework Convention on Climate Change, an international agreement that requires participating countries to develop and periodically submit an inventory of greenhouse gas emissions. Data and analysis for this indicator come from EPA's *Inventory of U.S. Greenhouse Gas Emissions and Sinks: 1990–2010*.[3] This indicator is restricted to emissions associated with human activities.

This indicator reports emissions of greenhouse gases according to their 100-year global warming potential, a measure of how much a given amount of the greenhouse gas is estimated to contribute to global warming over a period of 100 years after being emitted (see table on p. 10). For purposes of comparison, global warming potential values are calculated in relation to carbon dioxide and are expressed in terms of carbon dioxide equivalents. For additional perspective, this indicator also shows greenhouse gas emissions in relation to economic activity and population.

Figure 1. U.S. Greenhouse Gas Emissions by Gas, 1990–2010

This figure shows emissions of carbon dioxide, methane, nitrous oxide, and several fluorinated gases in the United States from 1990 to 2010. For consistency, emissions are expressed in million metric tons of carbon dioxide equivalents.

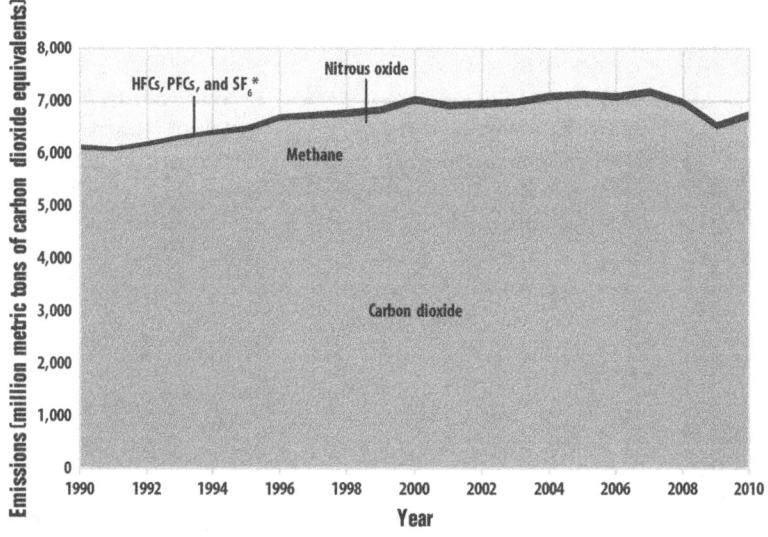

* HFCs are hydrofluorocarbons, PFCs are perfluorocarbons, and SF$_6$ is sulfur hexafluoride.

Data source: U.S. EPA, 2012[4]

Figure 2. U.S. Greenhouse Gas Emissions and Sinks by Economic Sector, 1990–2010

This figure shows greenhouse gas sinks (negative values) and emissions by source in the United States from 1990 to 2010. For consistency, emissions are expressed in million metric tons of carbon dioxide equivalents. Totals do not match Figure 1 exactly because the economic sectors shown here do not include emissions from U.S. territories.

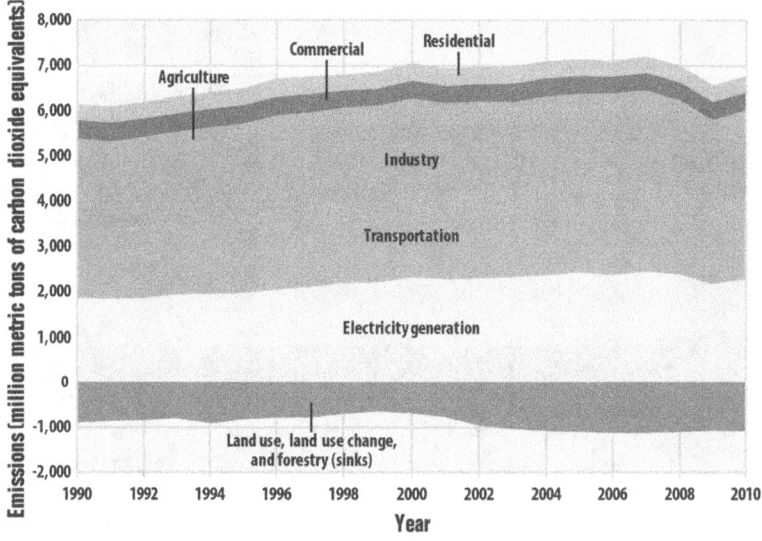

Data source: U.S. EPA, 2012[5]

Figure 3. U.S. Greenhouse Gas Emissions per Capita and per Dollar of GDP, 1990–2010

This figure shows trends in greenhouse gas emissions from 1990 to 2010 per capita (heavy orange line), based on the total U.S. population (thin orange line). It also shows trends in emissions compared with the real GDP (heavy blue line). Real GDP is the value of all goods and services produced in the country during a given year, adjusted for inflation (thin blue line). All data are indexed to 1990 as the base year, which is assigned a value of 100. For instance, a real GDP value of 163 in the year 2010 would represent a 63 percent increase since 1990.

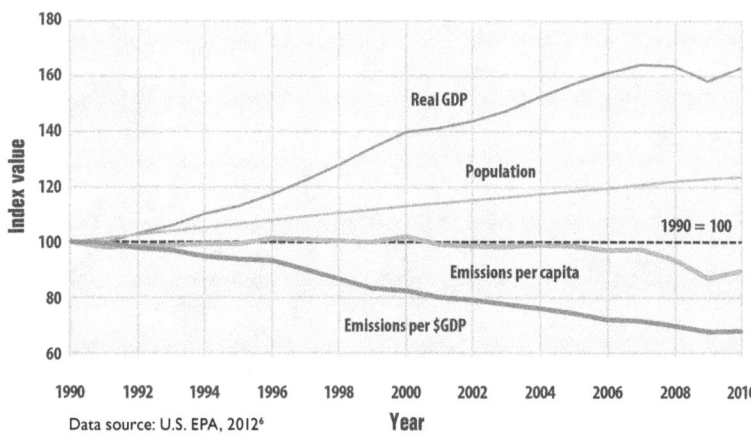

Data source: U.S. EPA, 2012[6]

Key Points

- In April 2010, U.S. greenhouse gas emissions totaled 6,822 million metric tons of carbon dioxide equivalents, a 10 percent increase from 1990 (see Figure 1).

- For the United States, during the period from 1990 to 2010 (see Figure 1):

  o Emissions of carbon dioxide, the primary greenhouse gas emitted by human activities, increased by 12 percent.

  o Methane emissions remained roughly the same, as higher emissions from activities such as livestock production and natural gas systems were largely offset by reduced emissions from landfills and coal mines.[7]

  o Nitrous oxide emissions, largely derived from vehicle emissions and agricultural soil management practices, such as the use of nitrogen as a fertilizer, declined by 3 percent.

  o Emissions of fluorinated gases (hydrofluorocarbons, perfluorocarbons, and sulfur hexafluoride), released as a result of commercial, industrial, and household uses, increased by 58 percent.

- Electricity generation is the largest U.S. emissions source, accounting for about 32 percent of total U.S. greenhouse gas emissions since 1990. Transportation is the second-largest source of greenhouse gas emissions, accounting for 27 percent of emissions since 1990 (see Figure 2).

- Emissions sinks, the opposite of emissions sources, absorb and store emissions. In 2010, 16 percent of U.S. greenhouse gas emissions were offset by sinks resulting from land use and forestry practices (see Figure 2). One major sink is the net growth of forests, which remove carbon from the atmosphere. Other carbon sinks are associated with how people use the land, including the practice of depositing yard trimmings and food scraps in landfills.

- Emissions increased at about the same rate as the population from 1990 to 2007, which caused emissions per capita to remain fairly level (see Figure 3). Total emissions and emissions per capita declined from 2007 to 2009, due in part to a drop in U.S. economic production during this time. Emissions have increased since 2009 as the U.S. economy has begun to grow again.[8]

- From 1990 to 2010, greenhouse gas emissions per dollar of U.S. gross domestic product (GDP) declined by 32 percent (see Figure 3). This change may reflect a combination of increased energy efficiency and structural changes in the economy.

Global Greenhouse Gas Emissions

Background

Since preindustrial times, increasing emissions of greenhouse gases due to human activities worldwide have led to a noticeable increase in atmospheric concentrations of long-lived and other greenhouse gases (see the Atmospheric Concentrations of Greenhouse Gases indicator on p. 16). Every country around the world emits greenhouse gases into the atmosphere, meaning the root causes of climate change are truly global. Some countries produce far more greenhouse gases than others, and several factors such as economic activity, population, income level, land use, and climatic conditions can influence a country's emissions levels. Tracking greenhouse gas emissions worldwide provides a global context for understanding the United States and other nations' roles in climate change.

About the Indicator

Like the U.S. Greenhouse Gas Emissions indicator (p. 12), this indicator focuses on emissions of gases covered under the United Nations Framework Convention on Climate Change: carbon dioxide, methane, nitrous oxide, and several fluorinated gases. These are all important greenhouse gases that are influenced by human activities, and the Convention requires participating countries to develop and periodically submit an inventory of emissions.

Data and analysis for this indicator come from the World Resources Institute's Climate Analysis Indicators Tool (CAIT), which compiles data from peer-reviewed and internationally recognized greenhouse gas inventories developed by EPA and other government agencies worldwide. Global estimates for carbon dioxide are published annually, but estimates for other gases, such as methane and nitrous oxide, are available only every fifth year.

(Continued on page 15)

Figure 1. Global Greenhouse Gas Emissions by Gas, 1990–2005

This figure shows worldwide emissions of carbon dioxide, methane, nitrous oxide, and several fluorinated gases from 1990 to 2005. For consistency, emissions are expressed in million metric tons of carbon dioxide equivalents. These totals do not include emissions due to land-use change or forestry.

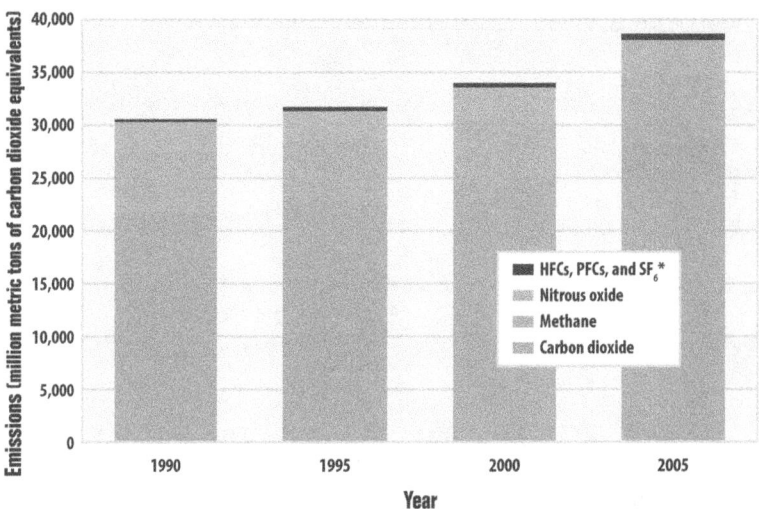

* HFCs are hydrofluorocarbons, PFCs are perfluorocarbons, and SF_6 is sulfur hexafluoride.

Data source: World Resources Institute, 2012[9]

Figure 2. Global Greenhouse Gas Emissions by Sector, 1990–2005

This figure shows worldwide greenhouse gas emissions by sector from 1990 to 2005. For consistency, emissions are expressed in million metric tons of carbon dioxide equivalents. These totals do not include emissions due to land-use change or forestry.*

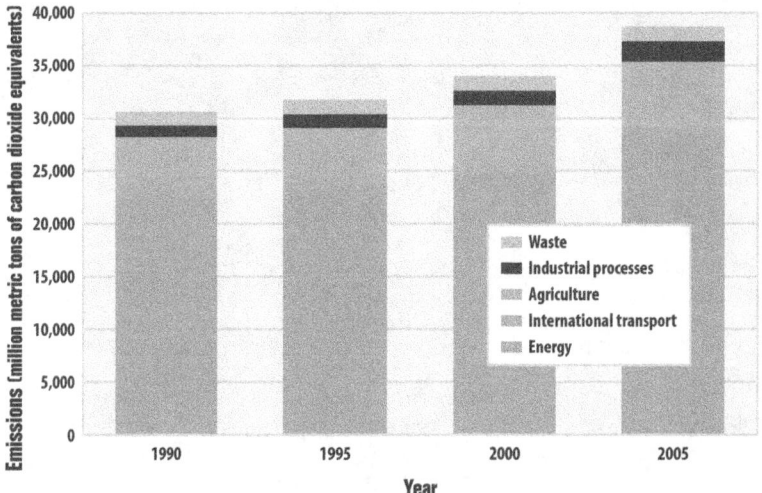

* Note that the sectors shown here are different from the economic sectors used in U.S. emissions accounting (see the U.S. Greenhouse Gas Emissions indicator). Emissions from international transport (aviation and marine) are separate from the energy sector because they are not part of individual countries' emissions inventories. The energy sector includes all other transportation activities.

Data source: World Resources Institute, 2012[10]

Key Points

- In 2005, estimated worldwide emissions totaled nearly 39 billion metric tons of greenhouse gases, expressed as carbon dioxide equivalents. This represents a 26 percent increase from 1990 (see Figures 1 and 2).

- Between 1990 and 2005, global emissions of all major greenhouse gases increased (see Figure 1). Emissions of carbon dioxide increased by 31 percent, which is particularly important because carbon dioxide accounts for nearly three-fourths of total global emissions. Methane emissions increased the least—10 percent—while emissions of nitrous oxide increased by 14 percent. Emissions of fluorinated gases more than doubled.

- Energy production and use (including energy used by vehicles) represent the largest source of greenhouse gas emissions worldwide (about 73 percent of the total), followed by agriculture (16 to 17 percent) (see Figure 2).

- Carbon dioxide emissions are increasing faster in some parts of the world than in others (see Figure 3).

Figure 3. Global Carbon Dioxide Emissions by Region, 1990–2008

This figure shows carbon dioxide emissions from 1990 to 2008 for different regions of the world. These totals do not include emissions due to land-use change or forestry.

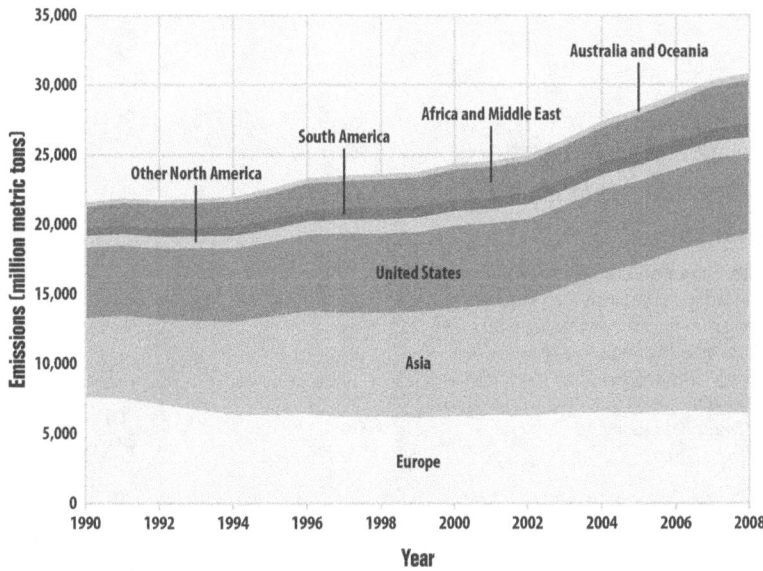

Data source: World Resources Institute, 2012[11]

This indicator tracks emissions of greenhouse gases according to their 100-year global warming potential, a measure of how much a given amount of the greenhouse gas is estimated to contribute to global warming over a period of 100 years after being emitted. For purposes of comparison, global warming potential values are calculated in relation to carbon dioxide and are expressed in terms of carbon dioxide equivalents.

Indicator Notes

Like the U.S. Greenhouse Gas Emissions indicator (p. 12), this indicator does not include emissions of a number of gases that affect climate but are not covered under the United Nations Framework Convention on Climate Change. For example, this indicator excludes ozone-depleting substances such as chlorofluorocarbons (CFCs) and hydrochlorofluorocarbons (HCFCs), which have high global warming potentials, because these gases are being phased out under an international agreement called the Montreal Protocol. This indicator is restricted to emissions associated with human activities, but it does not account for emissions associated with land-use change or forestry. There are also various emissions of greenhouse gases of natural origin, which this indicator does not cover.

Global emissions inventories for gases other than carbon dioxide are limited to five-year intervals. The United Nations Framework Convention on Climate Change database has more comprehensive data; however, these data are available mainly for a group of mostly developed countries that account for only about half of global greenhouse gas emissions. Thus, to provide a more representative measure of global greenhouse gas emissions, this indicator uses the broader CAIT database.

Data Sources

Data for this indicator came from the World Resources Institute's CAIT database, which is accessible online at: http://cait.wri.org. CAIT compiles data that were originally collected by organizations including the International Energy Agency, EPA, the U.S. Carbon Dioxide Information Analysis Center, and the European Commission.

Atmospheric Concentrations of Greenhouse Gases

This indicator describes how the levels of major greenhouse gases in the atmosphere have changed over time.

Background

Since the Industrial Revolution began in the late 1700s, people have added a significant amount of greenhouse gases into the atmosphere by burning fossil fuels, cutting down forests, and conducting other activities (see the U.S. and Global Greenhouse Gas Emissions indicators on pp. 12–15). When greenhouse gases are emitted into the atmosphere, many remain there for long time periods ranging from a decade to many millennia. Over time, these gases are removed from the atmosphere by emissions sinks, such as oceans, vegetation, or chemical reactions. Emissions sinks are the opposite of emissions sources, and they absorb and store emissions or cause the gases to break down. However, if these gases enter the atmosphere more quickly than they can be removed, their concentrations increase.

Many greenhouse gases remain in the atmosphere for decades or longer. The greenhouse gases being reported here become well mixed throughout the entire global atmosphere because of their long lifetimes and because of transport by winds. Concentrations of other greenhouse gases such as tropospheric ozone, which has an atmospheric lifetime of hours to days, often vary regionally and are not included in this indicator.

Concentrations of greenhouse gases are measured in parts per million (ppm), parts per billion (ppb), or parts per trillion (ppt) by volume. In other words, a concentration of 1 ppb for a given gas means there is one part of that gas in 1 billion parts of a given amount of air. For some greenhouse gases, even changes as small as a few parts per trillion can make a difference in global climate.

About the Indicator

This indicator describes concentrations of greenhouse gases in the atmosphere. It focuses on the major greenhouse gases that result from human activities. These include carbon dioxide, methane,

(Continued on page 18)

Figure 1. Global Atmospheric Concentrations of Carbon Dioxide Over Time

This figure shows concentrations of carbon dioxide in the atmosphere from hundreds of thousands of years ago through 2011. The data come from a variety of historical ice core studies and recent air monitoring sites around the world. Each line represents a different data source.

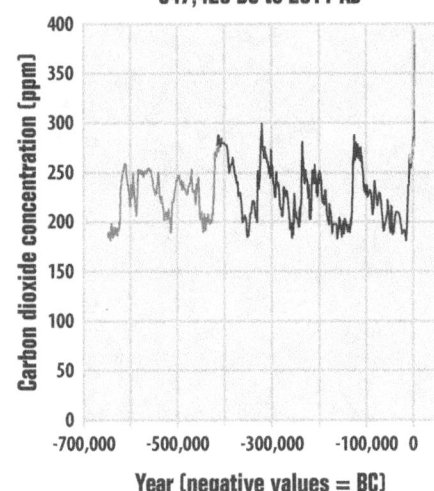

647,426 BC to 2011 AD

Data source: Various studies[12]

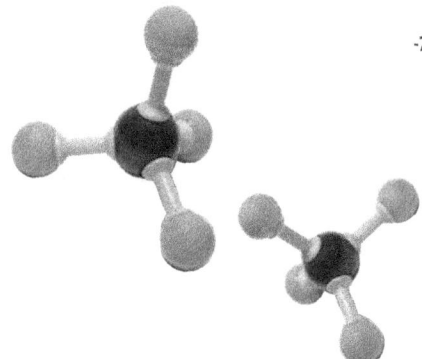

Figure 2. Global Atmospheric Concentrations of Methane Over Time

This figure shows concentrations of methane in the atmosphere from hundreds of thousands of years ago through 2011. The data come from a variety of historical ice core studies and recent air monitoring sites around the world. Each line represents a different data source.

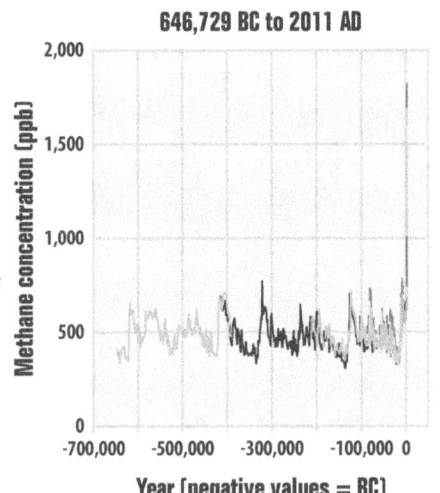

646,729 BC to 2011 AD

Data source: Various studies[13]

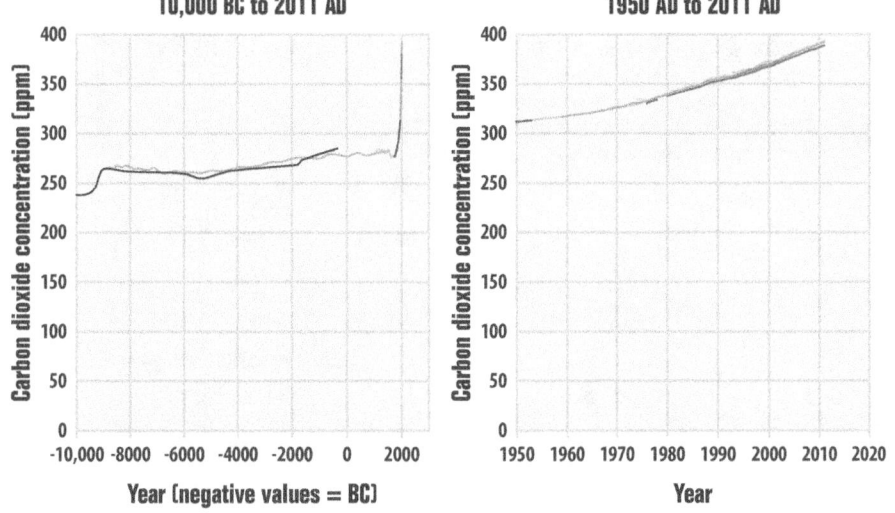

10,000 BC to 2011 AD

Carbon dioxide concentration [ppm] vs. Year (negative values = BC)

1950 AD to 2011 AD

Carbon dioxide concentration [ppm] vs. Year

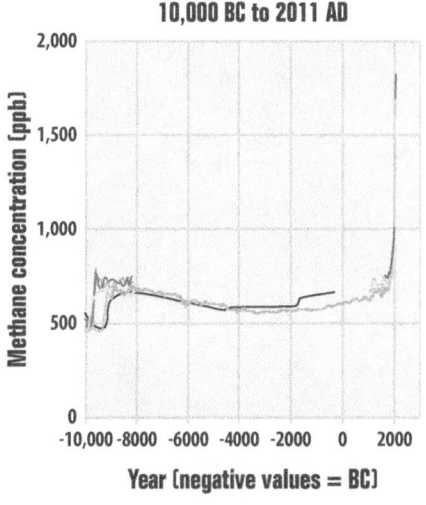

10,000 BC to 2011 AD

Methane concentration [ppb] vs. Year (negative values = BC)

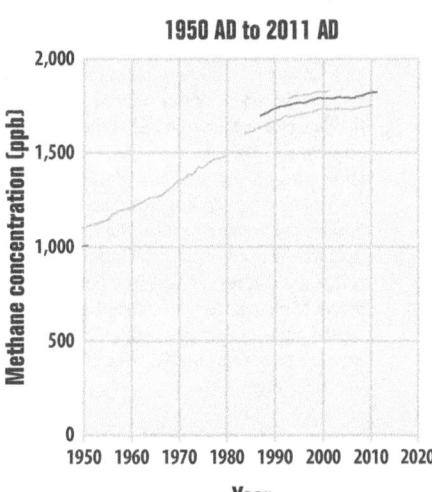

1950 AD to 2011 AD

Methane concentration [ppb] vs. Year

Key Points

- Global atmospheric concentrations of carbon dioxide, methane, nitrous oxide, and certain manufactured greenhouse gases have all risen over the last few hundred years (see Figures 1, 2, 3, and 4).

- Before the industrial era began in the late 1700s, carbon dioxide concentrations measured approximately 280 ppm. Concentrations have risen steadily since then, reaching 391 ppm in 2011—a 40 percent increase. Almost all of this increase is due to human activities.[14]

- The concentration of methane in the atmosphere has more than doubled since preindustrial times, reaching about 1,818 ppb in 2011. It is very likely that this increase is predominantly due to agriculture and fossil fuel use.[15]

- Historical measurements show that the current global atmospheric concentrations of carbon dioxide and methane are unprecedented compared with the past 650,000 years (see Figures 1 and 2).

- Over the past 100,000 years, concentrations of nitrous oxide in the atmosphere have rarely exceeded 280 ppb. Levels have risen since the 1920s, however, reaching a new high of 324 ppb in 2011 (see Figure 3). This increase is primarily due to agriculture.[16]

- Concentrations of many of the halogenated gases shown in Figure 4 (gases that contain chlorine, fluorine, or bromine) were essentially zero a few decades ago but have increased rapidly as they have been incorporated into industrial products and processes. Some of these chemicals are now being phased out of use because they are ozone-depleting substances, meaning they also cause harm to the Earth's ozone layer. As a result, concentrations of some ozone-depleting gases have begun to stabilize or decline (see Figure 4, left panel). Concentrations of other halogenated gases have continued to rise, however, especially where the gases have emerged as substitutes for ozone-depleting chemicals (see Figure 4, right panel). Some of these halogenated gases are considered major greenhouse gases due to their very high global warming potentials and long atmospheric lifetimes (see table on p. 10).

nitrous oxide, and certain manufactured gases known as halogenated gases. This indicator shows concentrations of greenhouse gases over thousands of years. Recent measurements come from monitoring stations around the world, while older measurements come from air bubbles trapped in layers of ice from Antarctica and Greenland. By determining the age of the ice layers and the concentrations of gases trapped inside, scientists can learn what the atmosphere was like thousands of years ago.

Indicator Notes

This indicator includes several of the most important halogenated gases, but some others are not shown. Many other halogenated gases are also greenhouse gases, but Figure 4 is limited to a set of common examples that represent most of the major types of these gases. The indicator also does not address certain other pollutants that can affect climate by either reflecting or absorbing energy. For example, sulfate particles can reflect sunlight away from the Earth, while black carbon aerosols (soot) absorb energy. Data for nitrogen trifluoride (Figure 4) reflect measurements made in the Northern Hemisphere only, where concentrations are expected to be slightly higher than the global average.

Data Sources

Global atmospheric concentration measurements for carbon dioxide (Figure 1), methane (Figure 2), and nitrous oxide (Figure 3) come from a variety of monitoring programs and studies published in peer-reviewed literature. References for the underlying data are included in the corresponding exhibits. Global atmospheric concentration data for selected halogenated gases (Figure 4) were compiled by the Advanced Global Atmospheric Gases Experiment,[17] the National Oceanic and Atmospheric Administration,[18] and two studies on nitrogen trifluoride.[19,20] An older figure with many of these gases appeared in the Intergovernmental Panel on Climate Change's Fourth Assessment Report.[21]

Figure 3. Global Atmospheric Concentrations of Nitrous Oxide Over Time

This figure shows concentrations of nitrous oxide in the atmosphere from 100,000 years ago through 2011. The data come from a variety of historical ice core studies and recent air monitoring sites around the world. Each line represents a different data source.

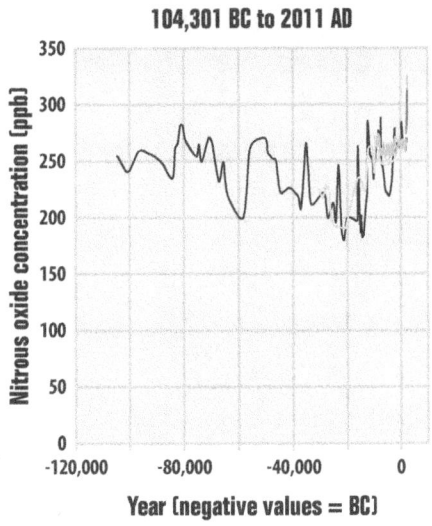

Data source: Various studies[22]

Figure 4. Global Atmospheric Concentrations of Selected Halogenated Gases, 1978–2011

This figure shows concentrations of several halogenated gases (which contain fluorine, chlorine, or bromine) in the atmosphere. The data come from monitoring sites around the world. Note that the scale is logarithmic, which means it increases by powers of 10. This is because the concentrations of different halogenated gases can vary by a few orders of magnitude. The numbers following the name of each gas (e.g., HCFC-22) are used to denote specific types of those particular gases.

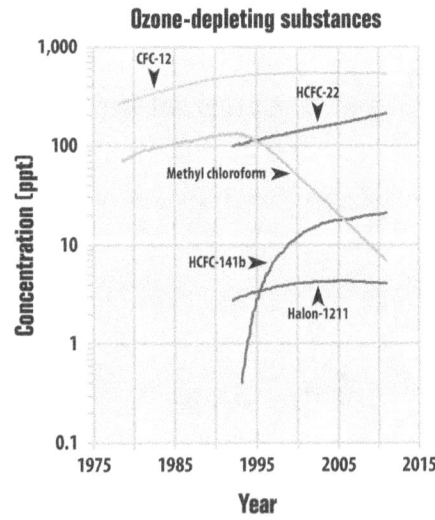

Data sources: AGAGE, 2011;[23] Arnold et al., 2012;[24] NOAA, 2011;[25] Weiss et al., 2008[26]

10,000 BC to 2011 AD

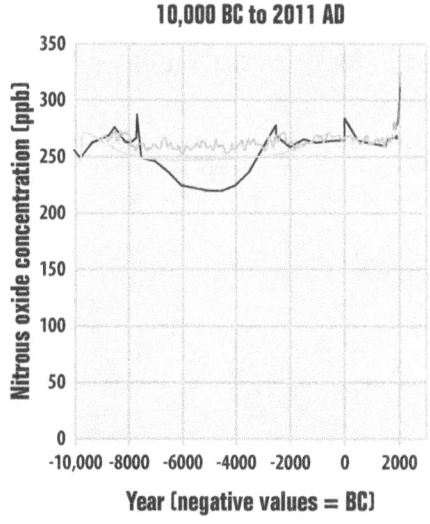

Year (negative values = BC)

1950 AD to 2011 AD

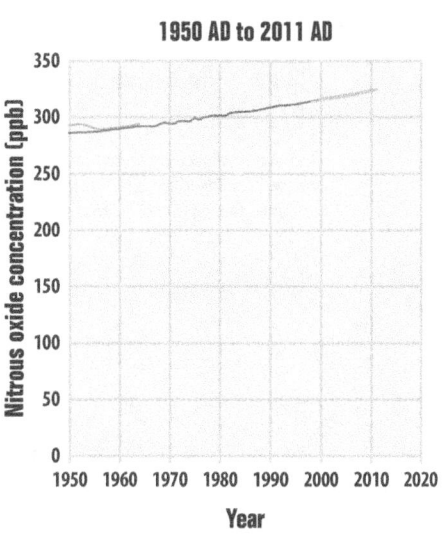

Year

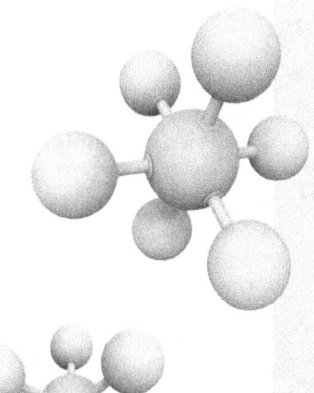

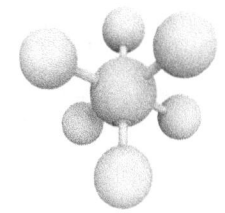

Other halogenated gases

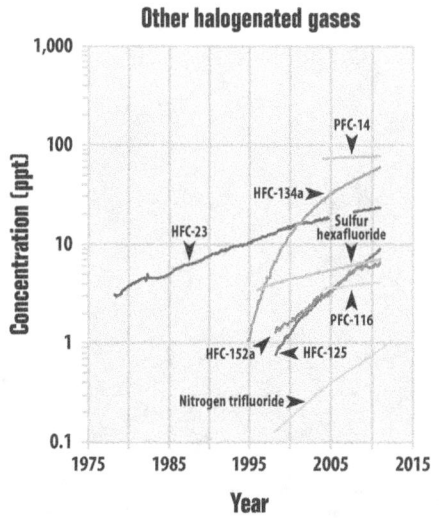

Year

Water Vapor as a Greenhouse Gas

Water vapor is the most abundant greenhouse gas in the atmosphere. Human activities have only a small direct influence on atmospheric concentrations of water vapor, primarily through irrigation and defor-estation, so it is not included in this indicator. However, the surface warming caused by human production of other greenhouse gases leads to an increase in atmospheric water vapor, because a warmer climate increases evaporation. This creates a positive "feedback loop" where warming leads to more warming.

Climate Forcing

This indicator measures the "radiative forcing" or heating effect caused by greenhouse gases in the atmosphere.

Background

When energy from the sun reaches the Earth, the planet absorbs some of this energy and radiates the rest back to space as heat. The Earth's surface temperature depends on this balance between incoming and outgoing energy. If this energy balance is altered, the Earth's average temperature will become warmer or cooler, leading to a variety of other changes in global climate.

A number of natural and human-influenced mechanisms can affect the global energy balance and force changes in the Earth's climate. Changes in greenhouse gas concentrations are one such mechanism. Greenhouse gases in the atmosphere absorb and re-emit most of the energy that radiates upward from the Earth's surface, adding the heat back to the lower atmosphere and warming the Earth's surface. Because elevated concentrations of many of the greenhouse gases emitted by human activities can remain in the atmosphere for decades, centuries, or longer, their associated warming effects persist over a long time. Factors that influence the Earth's energy balance can be quantified in terms of "radiative climate forcing." Positive radiative forcing indicates a warming influence (for example, by decreasing the amount of energy that escapes to space), while negative forcing is associated with a cooling influence. The balance between positive and negative forcing is what drives the actual change in surface temperature.

About the Indicator

This indicator measures the average total radiative forcing of 20 greenhouse gases, including carbon dioxide, methane, and nitrous oxide. The results were calculated by the National Oceanic and Atmospheric Administration based on measured concentrations of the gases in the atmosphere, compared with the concentrations that were present around 1750, before the Industrial Revolution began. Because each gas has a different capacity to absorb and emit heat energy, this indicator converts the changes in greenhouse gas

(Continued on page 21)

Figure 1. Radiative Forcing Caused by Major Greenhouse Gases, 1979–2011

This figure shows the amount of radiative forcing caused by various greenhouse gases, based on the concentrations present in the Earth's atmosphere. On the right side of the graph, radiative forcing has been converted to the Annual Greenhouse Gas Index, which is set to a value of 1.0 for 1990.

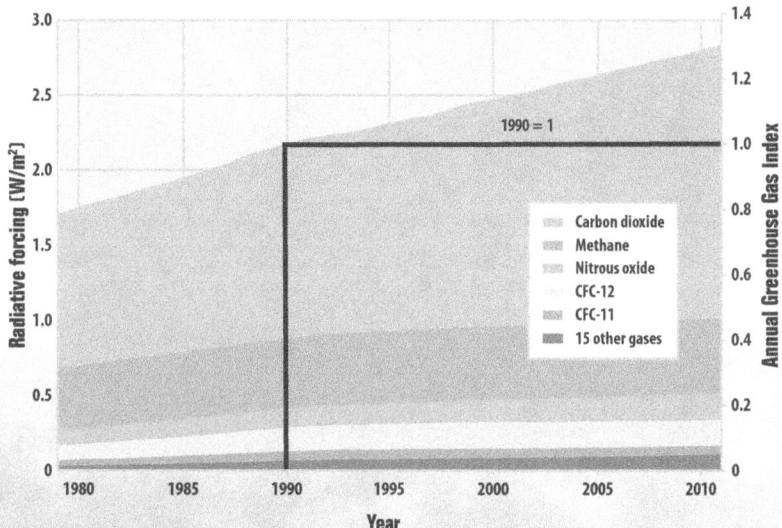

Data source: NOAA, 2012[27]

Key Points

- In 2011, the Annual Greenhouse Gas Index was 1.30, an increase in radiative forcing of 30 percent since 1990 (see Figure 1).

- Of the greenhouse gases shown in Figure 1, carbon dioxide accounts for by far the largest amount of radiative forcing, and its contribution continues to grow at a steady rate. By 2011, radiative forcing due to carbon dioxide was 40 percent higher than in 1990. Carbon dioxide accounts for approximately 80 percent of the overall increase in radiative forcing since 1990.

- Although the overall Annual Greenhouse Gas Index continues to rise, the rate of increase has slowed somewhat over time. This change has occurred in large part because methane concentrations have remained relatively steady since 1990 (although they have recently begun to rise again), and chlorofluorocarbon (CFC) concentrations have been declining because the production of these gases has been banned globally due to the harm they cause to the ozone layer (see Figure 1).

concentrations into a measure of the total radiative forcing (energy absorption) caused by each gas. Radiative forcing is calculated in watts per square meter, which represents the rate of energy transfer over a particular area.

The National Oceanic and Atmospheric Administration also translates the total radiative forcing of these measured gases into an index value called the Annual Greenhouse Gas Index. This number represents the ratio of the radiative forcing for a particular year compared with the radiative forcing in 1990, which is a common baseline year for global efforts to measure greenhouse gas concentrations. This indicator does not consider all substances that contribute to climate change (see Indicator Notes).

Indicator Notes

This indicator does not consider certain other substances that contribute to climate forcing. For example, the indicator does not measure reflective aerosol particles in the atmosphere, which can reduce radiative forcing, nor ground-level ozone or black carbon (soot), which can increase it. One gas shown in this indicator (methane) can also have an indirect influence on radiative forcing through its effects on water vapor and ozone formation; these indirect effects are not shown.

Data Sources

Data for this indicator were provided by the National Oceanic and Atmospheric Administration. This figure and other information are available at: www.esrl. noaa.gov/gmd/aggi.

Weather and

INDICATORS IN THIS CHAPTER

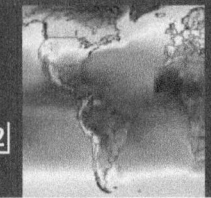

U.S. and Global Temperature

High and Low Temperatures

U.S. and Global Precipitation

Climate

Weather is the state of the atmosphere at any given time and place. Most of the weather that affects people, agriculture, and ecosystems takes place in the lower layer of the atmosphere, the troposphere (see diagram of the Earth's atmosphere at right). Familiar aspects of weather include temperature, precipitation, clouds, and wind. Severe weather conditions include hurricanes, tornadoes, blizzards, and droughts.

Climate is the long-term average of the weather in a given place. While the weather can change in minutes or hours, a change in climate is something that develops over longer periods of decades to centuries. Climate is defined not only by average temperature and precipitation, but also by the type, frequency, duration, and intensity of weather events such as heat waves, cold spells, storms, floods, and droughts. Weather can vary widely, and extreme events occur naturally, but average conditions tend to remain stable unless the Earth experiences a force that can shift the climate. At various times in the Earth's history, the climate has changed in response to forces such as large volcanic eruptions, changes in greenhouse gas concentrations, and shifts in the Earth's orbit around the sun.

What is happening?

The average temperature at the surface of the Earth has been increasing over the past century, primarily because human activities are adding large quantities of heat-trapping greenhouse gases to the atmosphere.

Unusually warm days and nights have also become more common in some places. Generally, warmer surface temperatures lead to an increase in evaporation from the oceans and land, leading to an increase in globally averaged precipitation. However, while some regions can get more precipitation, shifting storm patterns and increased evaporation can cause some areas to experience more severe droughts than they have in the past. Scientific studies also indicate that extreme weather events such as storms, floods, and hurricanes are likely to become more intense. However, because these extremes already vary naturally, it may be difficult over short time periods to distinguish whether changes in their intensity and frequency can be attributed to larger climate trends caused by human influences.

Why does it matter?

Climate variations can directly or indirectly affect many aspects of society—in both positive and disruptive ways. For example, warmer average temperatures reduce heating costs and improve conditions for growing some crops; yet extreme heat can increase illnesses and deaths among vulnerable populations and damage some crops. Precipitation can replenish water supplies and support agriculture, but intense storms can damage property, cause loss of life and population displacement, and temporarily disrupt essential services such as transportation, telecommunications, energy, and water supplies.

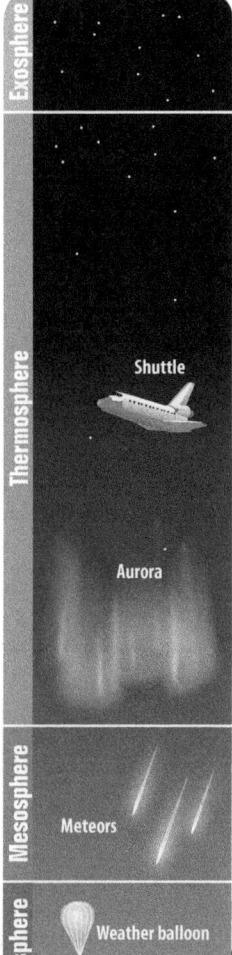

Earth's Atmosphere

Exosphere

Thermosphere — Shuttle — Aurora

Mesosphere — Meteors

Stratosphere — Weather balloon

Troposphere — Mount Everest

Source: NOAA, 2009[1]

For detailed information about data used in these indicators, see the online technical documentation at: www.epa.gov/climatechange/indicators.

 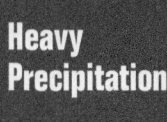 **Heavy Precipitation** **Drought**

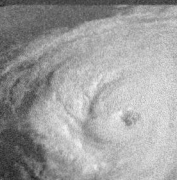 **Tropical Cyclone Activity**

U.S. and Global Temperature

This indicator describes trends in average surface temperature for the United States and the world.

Background

Temperature is a fundamental measurement for describing the climate, and the temperature in particular places can have wide-ranging effects on human life and ecosystems. For example, increases in air temperature can lead to more intense heat waves, which can cause illness and death, especially in vulnerable populations. Annual and seasonal temperature patterns also determine the types of animals and plants that can survive in particular locations. Changes in temperature can disrupt a wide range of natural processes, particularly if these changes occur more quickly than plant and animal species can adapt.

Concentrations of heat-trapping greenhouse gases are increasing in the Earth's atmosphere (see the Atmospheric Concentrations of Greenhouse Gases indicator on p. 16). In response, average temperatures at the Earth's surface are rising and are expected to continue rising. However, because climate change can shift the wind patterns and ocean currents that drive the world's climate system, some areas experience more warming than others, and some might experience cooling.

About the Indicator

This indicator examines U.S. and global surface temperature patterns from 1901 to the present. U.S. surface measurements come from weather stations on land, while global surface measurements also incorporate observations from buoys and ships on the ocean, thereby providing data from sites spanning much of the surface of the Earth. For comparison, this indicator also displays satellite measurements that can be used to estimate the temperature of the Earth's lower atmosphere since 1979.

This indicator shows anomalies, which compare recorded annual temperature values against a long-term average. For example, an anomaly of +2.0 degrees means the average temperature was 2 degrees higher than the long-term average. This indicator uses the average temperature from 1901 to 2000 as a

(Continued on page 25)

Figure 1. Temperatures in the Contiguous 48 States, 1901–2011

This figure shows how annual average temperatures in the contiguous 48 states have changed since 1901. Surface data come from land-based weather stations. Satellite measurements cover the lower troposphere, which is the lowest level of the Earth's atmosphere (see diagram on p. 23). "UAH" and "RSS" represent two different methods of analyzing the original satellite measurements. This graph uses the 1901 to 2000 average as a baseline for depicting change. Choosing a different baseline period would not change the shape of the data over time.

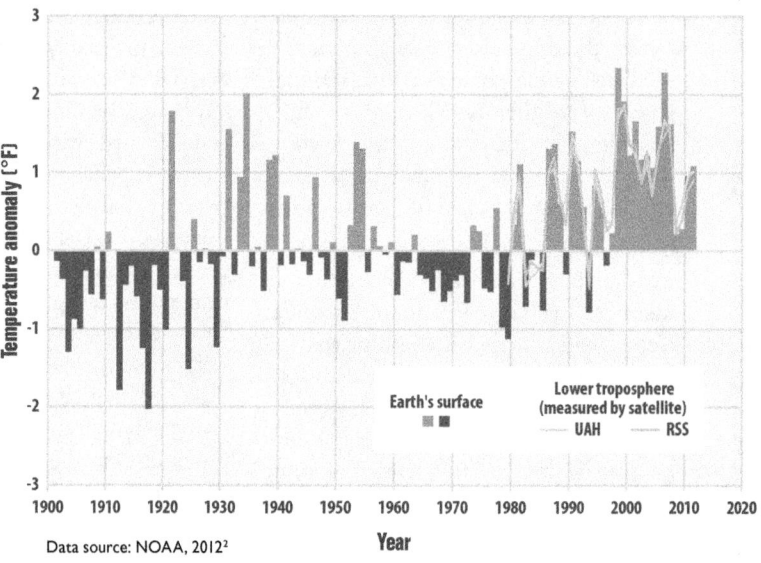

Data source: NOAA, 2012[2]

Figure 2. Temperatures Worldwide, 1901–2011

This figure shows how annual average temperatures worldwide have changed since 1901. Surface data come from a combined set of land-based weather stations and sea surface temperature measurements. Satellite measurements cover the lower troposphere, which is the lowest level of the Earth's atmosphere (see diagram on p. 23). "UAH" and "RSS" represent two different methods of analyzing the original satellite measurements. This graph uses the 1901 to 2000 average as a baseline for depicting change. Choosing a different baseline period would not change the shape of the data over time.

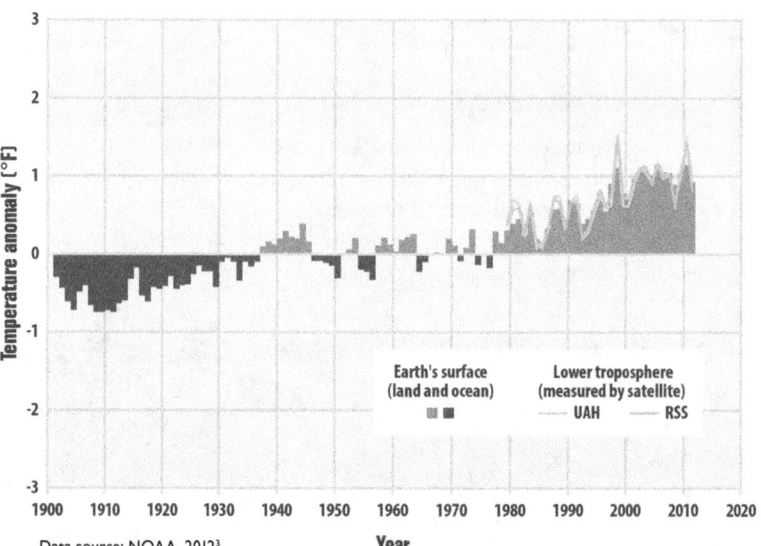

Data source: NOAA, 2012[3]

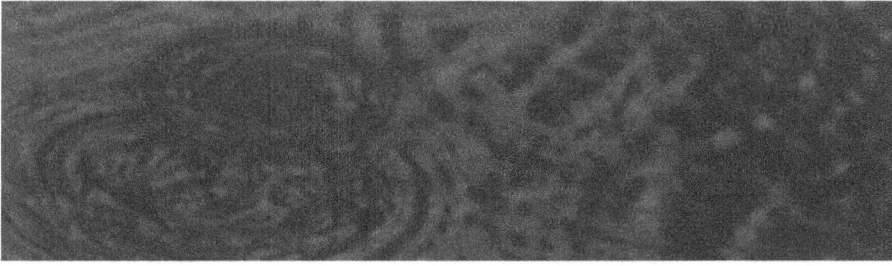

Figure 3. Rate of Temperature Change in the United States, 1901–2011

This figure shows how annual average air temperatures have changed in different parts of the United States since the early 20th century (since 1901 for the contiguous 48 states, 1905 for Hawaii, and 1918 for Alaska).

Rate of temperature change [°F per century]:

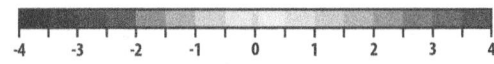

-4   -3   -2   -1   0   1   2   3   4

Data source: NOAA, 2012[4]   Gray interval: -0.1 to 0.1°F

baseline for comparison. Annual anomalies are calculated for each weather station, starting from daily and monthly average temperatures. Anomalies for broader regions have been determined by dividing the country (or the world) into a grid, averaging the data for all weather stations within each cell of the grid, and then averaging the grid cells together (for Figures 1 and 2) or displaying them on a map (Figure 3). This method ensures that the results are not biased toward regions that happen to have many stations close together.

Indicator Notes

Data from the early 20th century are somewhat less precise than more recent data because there were fewer stations collecting measurements at the time, especially in the Southern Hemisphere. However, the overall trends are still reliable. Where possible, the data have been adjusted to account for any biases that might be introduced by station moves, development (e.g., urbanization) near the station, changes in instruments and times of measurement, and other changes.

Data Sources

The data for this indicator were provided by the National Oceanic and Atmospheric Administration's National Climatic Data Center, which maintains a large collection of climate data online at: www.ncdc.noaa.gov/oa/ncdc.html. Surface temperature anomalies were calculated based on monthly values from a network of long-term monitoring stations. Satellite data were analyzed by two independent groups—the Global Hydrology and Climate Center at the University of Alabama in Huntsville (UAH) and Remote Sensing Systems (RSS)—resulting in slightly different trend lines.

Key Points

- Since 1901, the average surface temperature across the contiguous 48 states has risen at an average rate of 0.13°F per decade (1.3°F per century) (see Figure 1). Average temperatures have risen more quickly since the late 1970s (0.31 to 0.45°F per decade). Seven of the top 10 warmest years on record for the contiguous 48 states have occurred since 1990.

- Worldwide, 2001–2010 was the warmest decade on record since thermometer-based observations began. Global average surface temperature has risen at an average rate of 0.14°F per decade since 1901 (see Figure 2), similar to the rate of warming within the contiguous 48 states. Since the late 1970s, however, the United States has warmed faster than the global rate.

- Some parts of the United States have experienced more warming than others (see Figure 3). The North, the West, and Alaska have seen temperatures increase the most, while some parts of the Southeast have experienced little change. However, not all of these regional trends are statistically significant.

High and Low Temperatures

Background

Unusually hot or cold temperatures can result in prolonged extreme weather events like summer heat waves or winter cold spells. Heat waves can lead to illness and death, particularly among older adults, the very young, and other vulnerable groups (see the Heat-Related Deaths indicator on p. 72). People can also die from exposure to extreme cold (hypothermia). In addition, prolonged exposure to excessive heat and cold can damage crops and injure or kill livestock. Extreme heat can lead to power outages as heavy demands for air conditioning strain the power grid, while extremely cold weather increases the need for heating fuel.

Record-setting daily temperatures, heat waves, and cold spells are a natural part of day-to-day variation in weather. However, as the Earth's climate warms overall, heat waves are expected to become more frequent, longer, and more intense.[5,6] Higher heat index values (which combine temperature and humidity to describe perceived temperature) are expected to increase discomfort and could aggravate health issues. Conversely, cold spells are expected to decrease. In most locations, scientists expect daily minimum temperatures throughout the year to become warmer at a faster rate than daily maximum temperatures.[7]

About the Indicator

Trends in extreme temperatures can be examined in a variety of ways. This indicator covers several approaches by looking at prolonged heat wave events as well as unusually hot or cold daily highs and lows. The data come from thousands of weather stations across the United States. National patterns can be determined by dividing the country into a grid and examining the data for one station in each cell of the grid. This method ensures that the results are not biased toward regions that happen to have many stations close together.

Figure 1 shows the U.S. Annual Heat Wave Index, which tracks the occurrence of heat wave conditions across the contiguous 48 states from 1895 to 2011. While there is no universal definition of a heat wave, this index defines a heat wave as a four-day period with an average temperature that would only be expected to occur once every 10 years, based on the historical record. The index value

Figure 1. U.S. Annual Heat Wave Index, 1895–2011

This figure shows the annual values of the U.S. Heat Wave Index from 1895 to 2011. These data cover the contiguous 48 states.

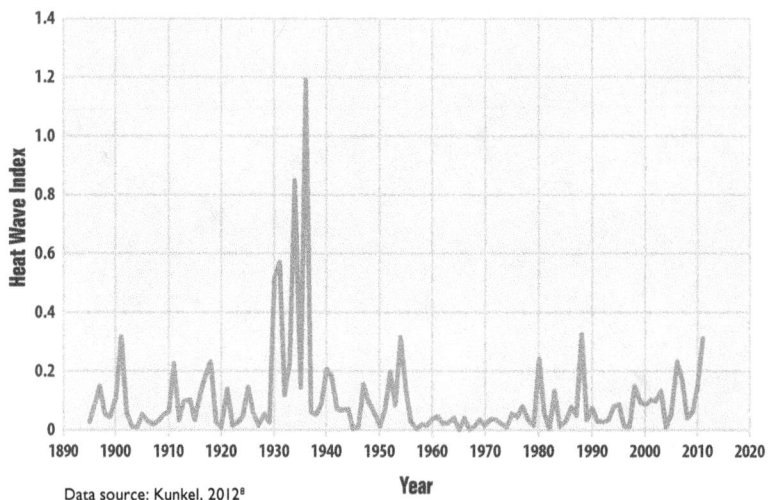

Data source: Kunkel, 2012[8]

Figure 2. Area of the Contiguous 48 States With Unusually Hot Summer Temperatures, 1910–2012

This graph shows the percentage of the land area of the contiguous 48 states with unusually hot daily high and low temperatures during the months of June, July, and August. The thin lines represent individual years, while the thick lines show a nine-year weighted average. Red lines represent daily highs, while orange lines represent daily lows. The term "unusual" is based on the long-term average conditions at each location.

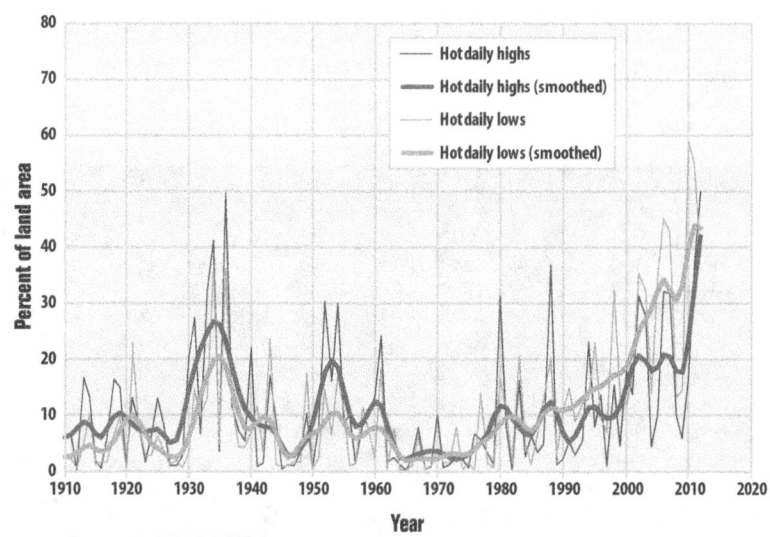

Data source: NOAA, 2012[9]

(Continued on page 27)

Figure 3. Area of the Contiguous 48 States With Unusually Cold Winter Temperatures, 1911–2012

This graph shows the percentage of the land area of the contiguous 48 states with unusually cold daily high and low temperatures during the months of December, January, and February. The thin lines represent individual years, while the thick lines show a nine-year weighted average. Blue lines represent daily highs, while purple lines represent daily lows. The term "unusual" is based on the long-term average conditions at each location.

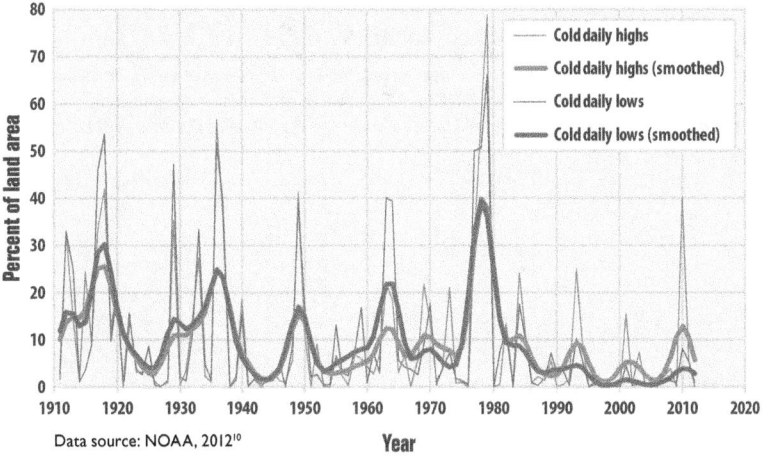

Data source: NOAA, 2012[10]

Figure 4. Record Daily High and Low Temperatures in the Contiguous 48 States, 1950–2009

This figure shows the percentage of daily temperature records set at weather stations across the contiguous 48 states by decade. Record highs (red) are compared with record lows (blue).

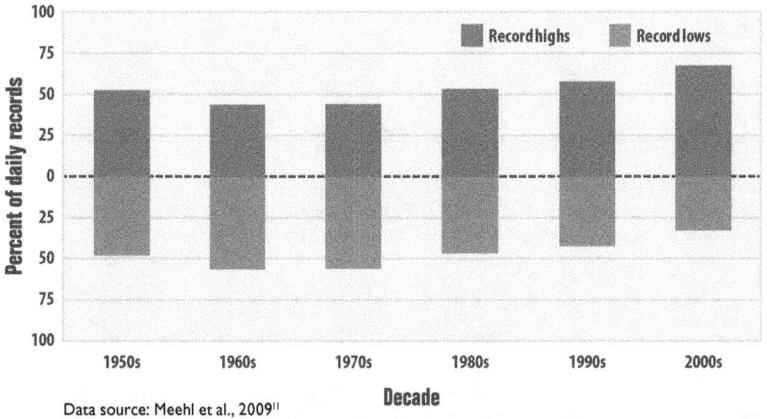

Data source: Meehl et al., 2009[11]

Key Points

- Heat waves were frequent and widespread in the 1930s, and these remain the most severe heat waves in the U.S. historical record (see Figure 1). Poor land use practices and many years of intense drought (the "Dust Bowl") contributed to these heat waves by depleting soil moisture and reducing the moderating effects of evaporation.[12]
- Unusually hot summer days (highs) have become more common over the last few decades (see Figure 2). The occurrence of unusually hot summer nights (lows) has increased at an even faster rate. This trend indicates less "cooling off" at night.
- The 20th century saw many winters with widespread patterns of unusually low temperatures, including a particularly large spike in the late 1970s (see Figure 3). Since the 1980s, though, unusually cold winter temperatures have become less common—particularly very cold nights (lows).
- If the climate were completely stable, one might expect to see highs and lows each accounting for about 50 percent of the records set. However, since the 1970s, record-setting daily high temperatures have become more common than record lows across the United States (see Figure 4). The most recent decade had twice as many record highs as record lows.

for a given year depends on how often heat waves occur and how widespread they are. For example, an index value of 0.2 could mean that 20 percent of the country experienced one heat wave, 10 percent of the country experienced two heat waves, or some other combination of frequency and area resulted in this value.

Figures 2 and 3 show trends in the percentage of the country's area experiencing unusually hot temperatures in the summer and unusually cold temperatures in the winter. These graphs are based on daily maximum temperatures, which usually occur during the day, and daily minimum temperatures, which usually occur at night. At each station, the recorded highs and lows are compared with the full set of historical records. After averaging over a particular month or season of interest, the coldest 10 percent of years are considered "unusually cold" and the warmest 10 percent are "unusually hot." For example, if last year's summer highs were the 10th warmest on record for a particular location with more than 100 years of data, that year's summer highs would be considered unusually warm. Data are available from 1910 to 2012 for summer (June through August) and from 1911 to 2012 for winter (December of the previous year through February).

Many people are familiar with record daily high and low temperatures, which are frequently mentioned in weather reports. Figure 4 looks at trends in these records by comparing the number of record-setting highs with the number of record-setting lows by decade. These data come from a set of weather stations that have collected data consistently from 1950 through 2009.

Indicator Notes

Temperature data are less certain for the early part of the 20th century because fewer stations were operating at that time. In addition, measuring devices and methods have changed over time, and some stations have moved. The data have been adjusted to the extent possible to account for some of these influences and biases, however, and these uncertainties are not sufficient to change the fundamental trends shown in the figures.

Data Sources

The data for this indicator are based on measurements from weather stations managed by the National Oceanic and Atmospheric Administration. Figure 1 uses data from the National Weather Service Cooperative Observer Network; these data are available online at: www.nws.noaa.gov/os/coop/what-is-coop.html. Figures 2 and 3 are based on the U.S. Climate Extremes Index; for data and a description of the index, see: www.ncdc.noaa.gov/extremes/cei.html. Figure 4 uses National Weather Service data processed by Meehl et al. (2009).[13]

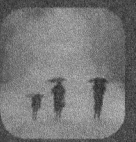

U.S. and Global Precipitation

Background

Precipitation can have wide-ranging effects on human well-being and ecosystems. Rainfall, snowfall, and the timing of snowmelt can all affect the amount of water available for drinking, irrigation, and industry, and can also determine what types of animals and plants (including crops) can survive in a particular place. Changes in precipitation can disrupt a wide range of natural processes, particularly if these changes occur more quickly than plant and animal species can adapt.

As average temperatures at the Earth's surface rise (see the U.S. and Global Temperature indicator on p. 24), more evaporation occurs, which, in turn, increases overall precipitation. Therefore, a warming climate is expected to increase precipitation in many areas. However, just as precipitation patterns vary across the world, so will the effects of climate change. By shifting the wind patterns and ocean currents that drive the world's climate system, climate change will also cause some areas to experience decreased precipitation. In addition, higher temperatures lead to more evaporation, so increased precipitation will not necessarily increase the amount of water available for drinking, irrigation, and industry (see the Drought indicator on p. 32).

About the Indicator

This indicator examines U.S. and global precipitation patterns from 1901 to the present, based on rainfall and snowfall measurements from land-based weather stations worldwide.

This indicator shows annual anomalies, or differences, compared with the average precipitation from 1901 to 2000. These anomalies are presented in terms of percent change compared with the baseline. Annual anomalies are calculated for each weather station. Anomalies for broader regions have been determined by dividing the country (or the world) into a grid, averaging the data for all weather stations within each cell of the grid, and then averaging the grid cells together (for Figures 1 and 2)

(Continued on page 29)

Figure 1. Precipitation in the Contiguous 48 States, 1901–2011

This figure shows how the total annual amount of precipitation in the contiguous 48 states has changed since 1901. This graph uses the 1901 to 2000 average as a baseline for depicting change. Choosing a different baseline period would not change the shape of the data over time.

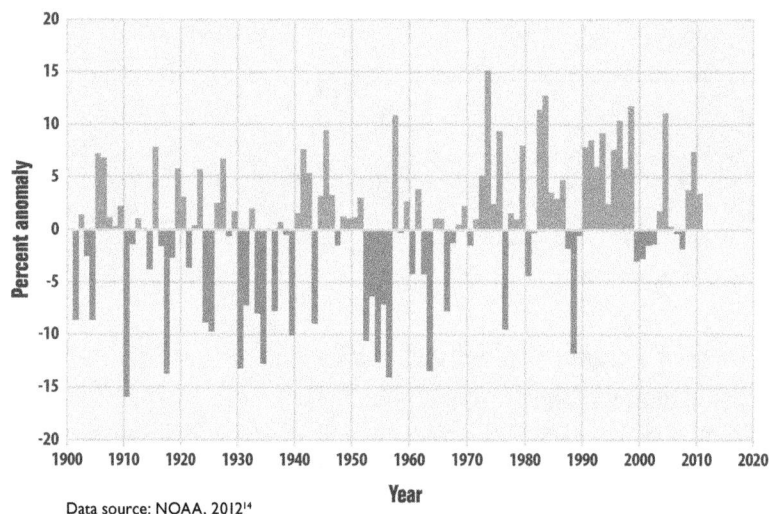

Data source: NOAA, 2012[14]

Figure 2. Precipitation Worldwide, 1901–2011

This figure shows how the total annual amount of precipitation over land worldwide has changed since 1901. This graph uses the 1901 to 2000 average as a baseline for depicting change. Choosing a different baseline period would not change the shape of the data over time.

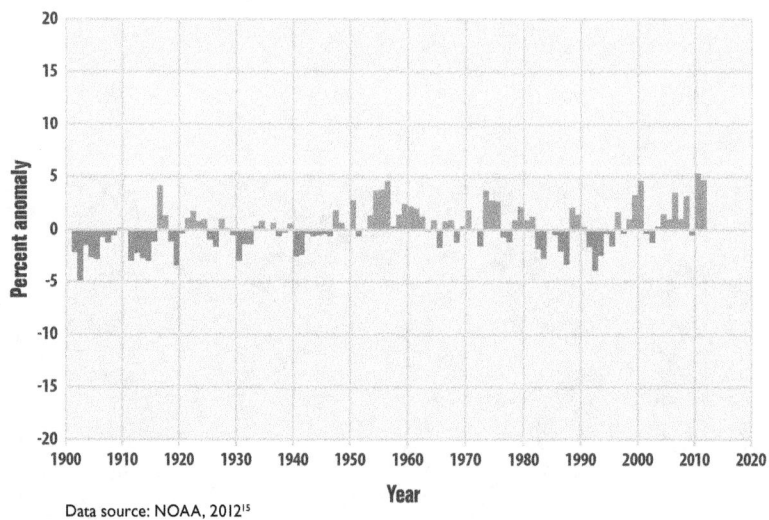

Data source: NOAA, 2012[15]

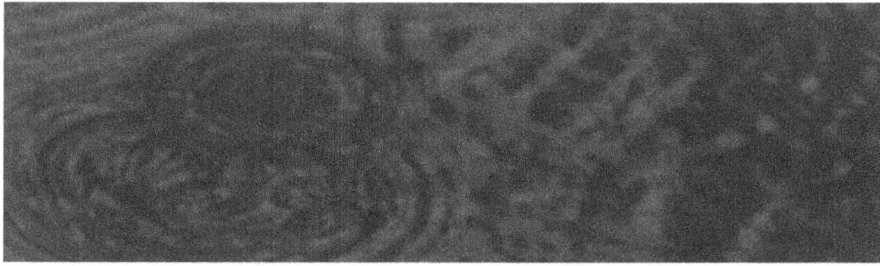

Figure 3. Rate of Precipitation Change in the United States, 1901–2011

This figure shows the rate of change in total annual precipitation in different parts of the United States since the early 20ᵗʰ century (since 1901 for the contiguous 48 states, 1905 for Hawaii, and 1918 for Alaska).

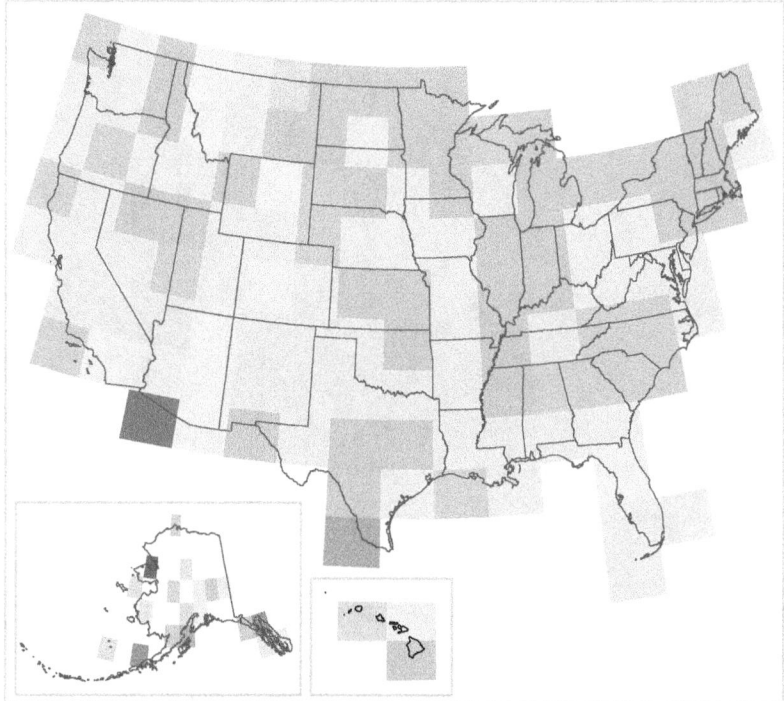

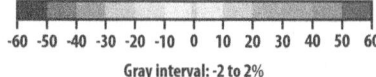

Rate of change in precipitation (% per century):

-60 -50 -40 -30 -20 -10 0 10 20 30 40 50 60

Gray interval: -2 to 2%

Data source: NOAA, 2012[16]

or displaying them on a map (Figure 3). This method ensures that the results are not biased toward regions that happen to have many stations close together.

Indicator Notes

Data from the early 20ᵗʰ century are somewhat less precise because there were fewer stations collecting measurements at the time. To ensure that overall trends are reliable, the data have been adjusted where possible to account for any biases that might be introduced by station moves, changes in measurement instruments, and other changes.

Data Sources

The data for this indicator were provided by the National Oceanic and Atmospheric Administration's National Climatic Data Center, which maintains a large collection of climate data online at: www.ncdc.noaa. gov/oa/ncdc.html. Global, U.S., and regional precipitation anomalies were calculated based on monthly values from a network of long-term monitoring stations.

Key Points

- On average, total annual precipitation has increased over land areas in the United States and worldwide (see Figures 1 and 2). Since 1901, global precipitation has increased at an average rate of 2.3 percent per century, while precipitation in the contiguous 48 states has increased at a rate of 5.9 percent per century.

- Some parts of the United States have experienced greater increases in precipitation than others. A few areas such as Hawaii and parts of the Southwest have seen a decrease in precipitation (see Figure 3).

Heavy Precipitation

Background

"Heavy precipitation" refers to instances during which the amount of precipitation experienced in a location substantially exceeds what is normal. What constitutes a period of heavy precipitation varies according to location and season.

Climate change can affect the intensity and frequency of precipitation. Warmer oceans increase the amount of water that evaporates into the air. When more moisture-laden air moves over land or converges into a storm system, it can produce more intense precipitation—for example, heavier rain and snow storms.[17] The potential impacts of heavy precipitation include crop damage, soil erosion, and an increase in flood risk due to heavy rains. In addition, runoff from precipitation can impair water quality as pollutants deposited on land wash into water bodies.

Heavy precipitation does not necessarily mean the total amount of precipitation at a location has increased—just that precipitation is occurring in more intense events. However, changes in the intensity of precipitation, when combined with changes in the interval between precipitation events, can also lead to changes in overall precipitation totals.

About the Indicator

Heavy precipitation events can be measured by tracking their frequency, examining their return period (the chance that the event will be equaled or exceeded in a given year), or directly measuring the amount of precipitation in a certain period.

One way to track heavy precipitation is by calculating what percentage of a particular location's total precipitation in a given year has come in the form of extreme one-day events—or, in other words, what percentage of precipitation is arriving in short, intense bursts. Figure 1 of this indicator looks at the prevalence of extreme single-day precipitation events over time.

(Continued on page 31)

Figure 1. Extreme One-Day Precipitation Events in the Contiguous 48 States, 1910–2011

This figure shows the percentage of the land area of the contiguous 48 states where a much greater than normal portion of total annual precipitation has come from extreme single-day precipitation events. The bars represent individual years, while the line is a nine-year weighted average.

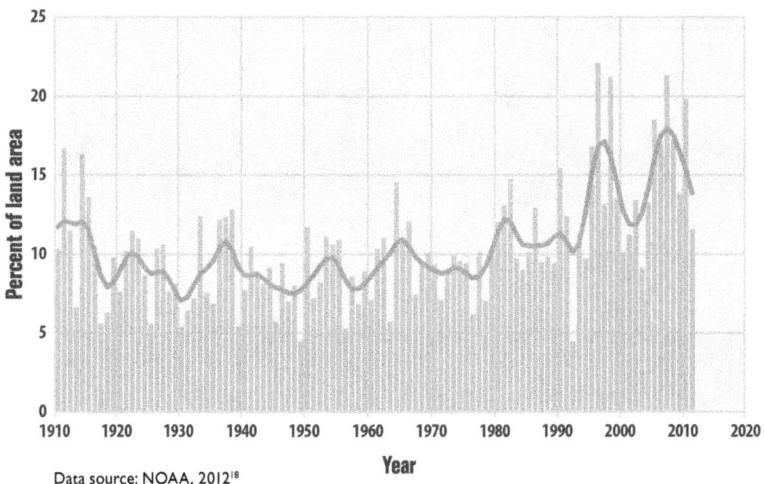

Data source: NOAA, 2012[18]

Figure 2. Unusually High Annual Precipitation in the Contiguous 48 States, 1895–2011

This figure shows the percentage of the land area of the contiguous 48 states that experienced much greater than normal precipitation in any given year, which means it scored 2.0 or above on the annual Standardized Precipitation Index (SPI). The thicker orange line shows a nine-year weighted average that smoothes out some of the year-to-year fluctuations.

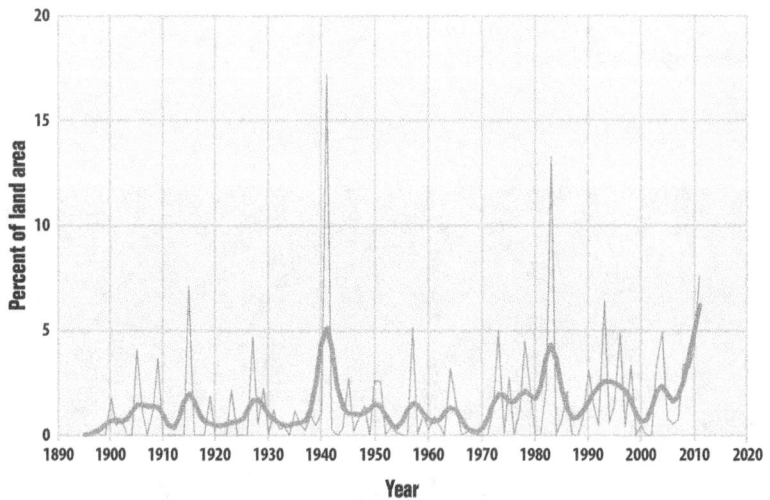

Data source: NOAA, 2012[19]

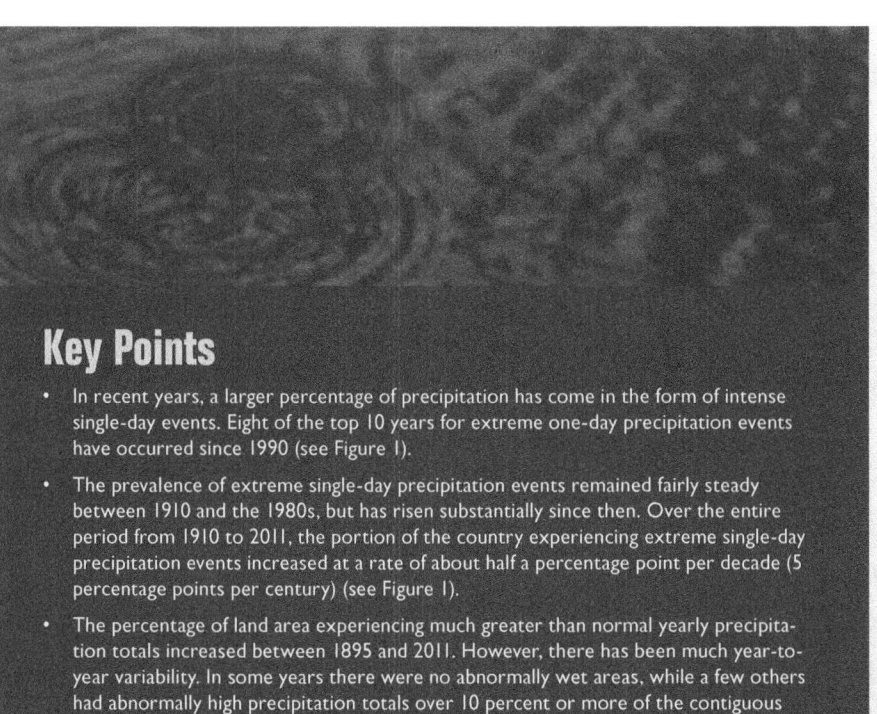

Key Points

- In recent years, a larger percentage of precipitation has come in the form of intense single-day events. Eight of the top 10 years for extreme one-day precipitation events have occurred since 1990 (see Figure 1).

- The prevalence of extreme single-day precipitation events remained fairly steady between 1910 and the 1980s, but has risen substantially since then. Over the entire period from 1910 to 2011, the portion of the country experiencing extreme single-day precipitation events increased at a rate of about half a percentage point per decade (5 percentage points per century) (see Figure 1).

- The percentage of land area experiencing much greater than normal yearly precipitation totals increased between 1895 and 2011. However, there has been much year-to-year variability. In some years there were no abnormally wet areas, while a few others had abnormally high precipitation totals over 10 percent or more of the contiguous 48 states' land area (see Figure 2). For example, 1941 was extremely wet in the West, while 1982 was very wet nationwide.[20]

- Figures 1 and 2 are both consistent with other studies that have found an increase in heavy precipitation over timeframes ranging from single days to 90-day periods to whole years.[21] For more information on trends in overall precipitation levels, see the U.S. and Global Precipitation indicator on p. 28.

For added insight, this indicator also tracks the occurrence of unusually high total yearly precipitation. It does so by looking at the Standardized Precipitation Index (SPI), which compares actual yearly precipitation totals with the range of precipitation totals that one would typically expect at a specific location, based on historical data. If a location experiences less precipitation than normal during a particular period, it will receive a negative SPI score, while a period with more precipitation than normal will receive a positive score. The more precipitation (compared with normal), the higher the SPI score. The SPI is a useful way to look at precipitation totals because it allows comparison of different locations and different seasons on a standard scale. Figure 2 shows what percentage of the total area of the contiguous 48 states had an annual SPI score of 2.0 or above (well above normal) in any given year.

Indicator Notes

Weather monitoring stations tend to be closer together in the eastern and central states than in the western states. In areas with fewer monitoring stations, heavy precipitation indicators are less likely to reflect local conditions accurately.

Data Sources

The data used for this indicator come from a large national network of weather stations and were provided by the National Oceanic and Atmospheric Administration's National Climatic Data Center. Figure 1 is based on Step #4 of the National Oceanic and Atmospheric Administration's U.S. Climate Extremes Index; for data and a description of the index, see: www.ncdc.noaa.gov/extremes/cei.html. Figure 2 is based on the U.S. SPI, which is shown in a variety of maps available online at: www.ncdc.noaa.gov/oa/climate/research/prelim/drought/spi.html. The data used to construct these maps are available from the National Oceanic and Atmospheric Administration at: ftp://ftp.ncdc.noaa.gov/pub/data/cirs.

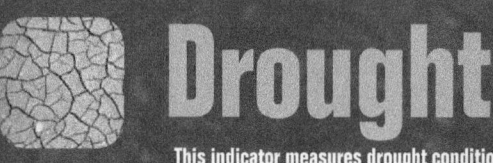

Drought

This indicator measures drought conditions of U.S. lands.

Background

There are many definitions and types of drought. Meteorologists generally define drought as a prolonged period of dry weather caused by a lack of precipitation that results in a serious water shortage for some activity, population, or ecological system. Drought can also be thought of as an extended imbalance between precipitation and evaporation.

As average temperatures have risen because of climate change, the Earth's water cycle has sped up through an increase in the rate of evaporation. An increase in evaporation makes more water available in the air for precipitation, but contributes to drying over some land areas, leaving less moisture in the soil. Thus, as the climate continues to change, many areas are likely to experience increased precipitation (see the U.S. and Global Precipitation indicator on p. 28) and increased risk of flooding (see the Heavy Precipitation indicator on p. 30), while areas located far from storm tracks are likely to experience less precipitation and increased risk of drought. As a result, since the 1950s, some regions of the world have experienced longer and more intense droughts, particularly in southern Europe and West Africa, while other regions have seen droughts become less frequent, less intense, or shorter (for example, in central North America).[22]

Drought conditions can negatively affect agriculture, water supplies, energy production, and many other aspects of society. The impacts vary depending on the type, location, intensity, and duration of the drought. For example, effects on agriculture can range from slowed plant growth to severe crop losses, while water supply impacts can range from lowered reservoir levels and dried-up streams to major water shortages. Lower streamflow and ground water levels can also harm plants and animals, and dried-out vegetation increases the risk of wildfires.

About the Indicator

During the 20th century, many indices were created to measure drought severity by looking at precipitation, soil moisture, stream flow, vegetation health, and other variables.[23] Figure 1 shows annual values of the most widely used index, the Palmer

(Continued on page 33)

Key Points

- Average drought conditions across the nation have varied since records began in 1895. The 1930s and 1950s saw the most widespread droughts, while the last 50 years have generally been wetter than average (see Figure 1).

- Over the period from 2000 through 2011, roughly 30 to 60 percent of the U.S. land area experienced conditions that were at least abnormally dry at any given time (see Figure 2). The years 2002, 2003, and 2007 were relatively high drought years, while 2001, 2005, 2009, and 2010 were relatively low drought years. In 2011, the overall extent of drought was relatively low, but the droughts that did occur were more severe than at any other time since 2000.

- According to the Drought Monitor, more than 64 percent of the contiguous U.S. land area was covered by moderate or greater drought by the end of September 2012. In many portions of the country, 2012 has been among the driest years on record.[24]

Figure 1. Average Drought Conditions in the Contiguous 48 States, 1895–2011

This chart shows annual values of the Palmer Drought Severity Index, averaged over the entire area of the contiguous 48 states. Positive values represent wetter-than-average conditions, while negative values represent drier-than-average conditions. A value between -2 and -3 indicates moderate drought, -3 to -4 is severe drought, and -4 or below indicates extreme drought. The thicker line is a nine-year weighted average.

Data source: NOAA, 2012[25]

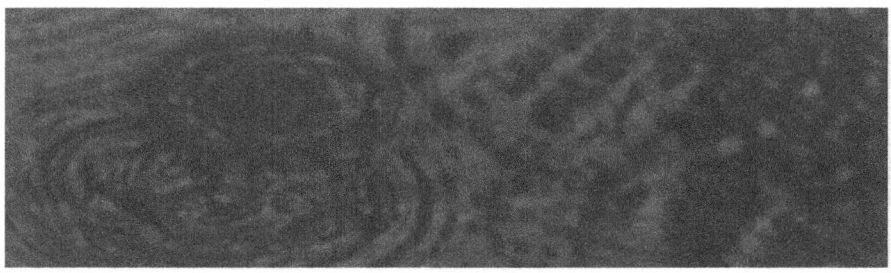

Figure 2. U.S. Lands Under Drought Conditions, 2000–2011

This chart shows the percentage of U.S. lands classified under drought conditions from 2000 through 2011. This figure uses the U.S. Drought Monitor classification system, which is described in the table below. The data cover all 50 states plus Puerto Rico.

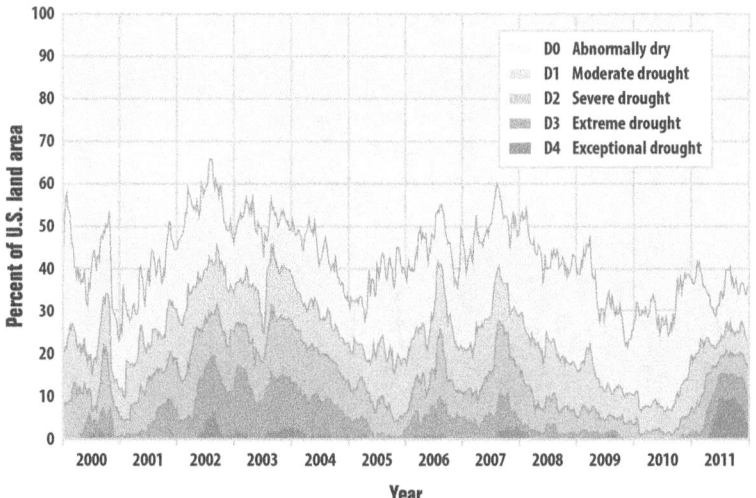

Data source: National Drought Mitigation Center, 2012[26]

Categories of Drought Severity

Category	Description	Possible Impacts
D0	Abnormally dry	Going into drought: short-term dryness slowing planting or growth of crops or pastures. Coming out of drought: some lingering water deficits; pastures or crops not fully recovered.
D1	Moderate drought	Some damage to crops or pastures; streams, reservoirs, or wells low; some water shortages developing or imminent; voluntary water use restrictions requested.
D2	Severe drought	Crop or pasture losses likely; water shortages common; water restrictions imposed.
D3	Extreme drought	Major crop/pasture losses; widespread water shortages or restrictions
D4	Exceptional drought	Exceptional and widespread crop/pasture losses; shortages of water in reservoirs, streams, and wells, creating water emergencies.

Experts update the U.S. Drought Monitor weekly and produce maps that illustrate current conditions as well as short- and long-term trends. Major participants include the National Oceanic and Atmospheric Administration, the U.S. Department of Agriculture, and the National Drought Mitigation Center. For a map of current drought conditions, visit the Drought Monitor website at: http://droughtmonitor.unl.edu.

Drought Severity Index, which is calculated from precipitation and temperature measurements at weather stations. An index value of zero represents average moisture conditions for a given location, based on many years of observations. A positive value means conditions are wetter than average, while a negative value is drier than average. Index values from locations across the contiguous 48 states have been averaged together to produce the national values shown in Figure 1.

For a more detailed perspective on recent trends, Figure 2 shows a newer index called the Drought Monitor, which is based on several indices (including Palmer), along with additional factors such as snow water content, ground water levels, reservoir storage, pasture/range conditions, and other impacts. The Drought Monitor uses codes from D0 to D4 (see table below) to classify drought severity. This part of the indicator covers all 50 states and Puerto Rico.

Indicator Notes

Because this indicator focuses on national trends, it does not show how drought conditions vary by region. For example, even if half of the country suffered from severe drought, Figure 1 could show an average index value close to zero if the rest of the country was wetter than average. Thus, Figure 1 might understate the degree to which droughts are becoming more severe in some areas while other places receive more rain as a result of climate change.

The U.S. Drought Monitor (Figure 2) offers a closer look at the percentage of the country that is affected by drought. However, this index is relatively new and thus too short-lived to be used for assessing long-term climate trends. With several decades of data collection, future versions of this indicator should be able to paint a more complete picture of trends over time.

Overall, this indicator gives a broad overview of drought conditions in the United States. It is not intended to replace local or state information that might describe conditions more precisely for a particular region.

Data Sources

Data for Figure 1 were obtained from the National Oceanic and Atmospheric Administration's National Climatic Data Center, which maintains a large collection of climate data online at: www.ncdc.noaa.gov/oa/ncdc.html. Data for Figure 2 were provided by the National Drought Mitigation Center. Historical data in table form are available at: http://droughtmonitor.unl.edu/archive.html. Maps and current drought information can be found on the main Drought Monitor site at: http://droughtmonitor.unl.edu.

Tropical Cyclone Activity

This indicator examines the frequency, intensity, and duration of hurricanes and other tropical storms in the Atlantic Ocean, Caribbean, and Gulf of Mexico.

Background

Hurricanes, tropical storms, and other intense rotating storms fall into a general category called cyclones. There are two main types of cyclones: tropical and extratropical (those that form outside the tropics). Tropical cyclones get their energy from warm tropical oceans. Extratropical cyclones get their energy from the jet stream and from temperature differences between cold, dry air masses from higher latitudes and warm, moist air masses from lower latitudes.

This indicator focuses on tropical cyclones in the Atlantic Ocean, Caribbean, and Gulf of Mexico. Tropical cyclones are most common during the "hurricane season," which runs from June through November. The effects of tropical cyclones are numerous and well known. At sea, storms disrupt and endanger shipping traffic. When cyclones encounter land, their intense rains and high winds can cause severe property damage, loss of life, soil erosion, and flooding. The associated storm surge—the large volume of ocean water pushed toward shore by the cyclone's strong winds—can cause severe flooding and destruction.

Climate change is expected to affect tropical cyclones by increasing sea surface temperatures, a key factor that influences cyclone formation and behavior. According to the U.S. Global Change Research Program, it is very likely that increased levels of greenhouse gases have contributed to an increase in sea surface temperatures in areas where hurricanes form.[27] The U.S. Global Change Research Program and the Intergovernmental Panel on Climate Change project that tropical cyclones will likely become more intense over the 21st century, with higher wind speeds and heavier rains.[28,29]

About the Indicator

Records of tropical cyclones in the Atlantic Ocean have been collected since the 1800s. The most reliable long-term records focus on hurricanes, which are the strongest category of tropical cyclones in the Atlantic, with wind speeds of at least 74 miles per hour. This indicator uses historical data from the National Oceanic and Atmospheric Administration to track the number of hurricanes per year in the North Atlantic (north of the equator) and the number reaching the United States since 1878. Some hurricanes over the ocean might have been

(Continued on page 35)

Figure 1. Number of Hurricanes in the North Atlantic, 1878–2011

This graph shows the number of hurricanes that formed in the North Atlantic Ocean each year from 1878 to 2011, along with the number that made landfall in the United States. The blue curve shows how the total count in the red curve can be adjusted to attempt to account for the lack of aircraft and satellite observations in early years. All three curves have been smoothed using a five-year average, plotted at the middle year. The most recent average (2007–2011) is plotted at 2009.

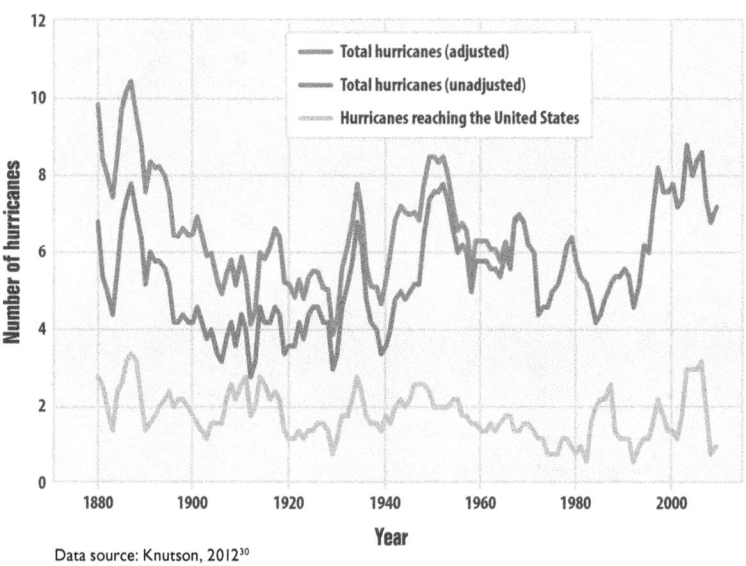

Data source: Knutson, 2012[30]

Figure 2. North Atlantic Tropical Cyclone Activity According to the Accumulated Cyclone Energy Index, 1950–2011

This figure shows total annual Accumulated Cyclone Energy (ACE) Index values from 1950 through 2011. The National Oceanic and Atmospheric Administration has defined "near normal," "above normal," and "below normal" ranges based on the distribution of ACE Index values over the 30 years from 1981 to 2010.

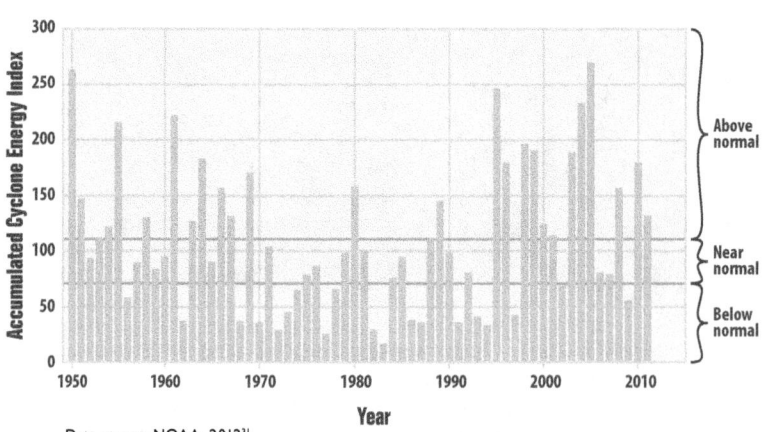

Data source: NOAA, 2012[31]

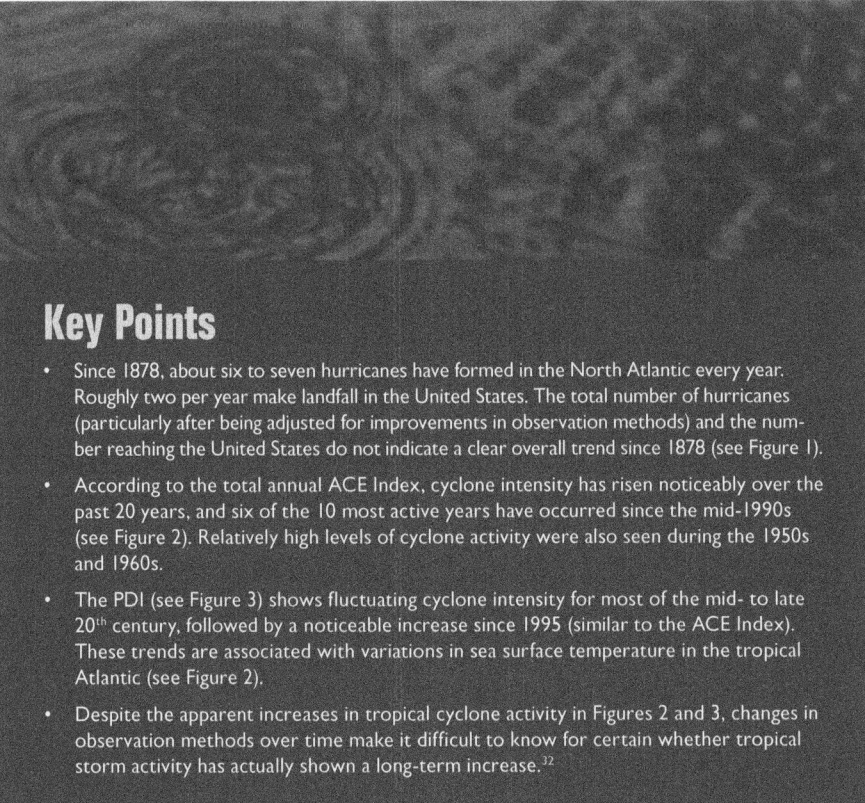

Key Points

- Since 1878, about six to seven hurricanes have formed in the North Atlantic every year. Roughly two per year make landfall in the United States. The total number of hurricanes (particularly after being adjusted for improvements in observation methods) and the number reaching the United States do not indicate a clear overall trend since 1878 (see Figure 1).

- According to the total annual ACE Index, cyclone intensity has risen noticeably over the past 20 years, and six of the 10 most active years have occurred since the mid-1990s (see Figure 2). Relatively high levels of cyclone activity were also seen during the 1950s and 1960s.

- The PDI (see Figure 3) shows fluctuating cyclone intensity for most of the mid- to late 20th century, followed by a noticeable increase since 1995 (similar to the ACE Index). These trends are associated with variations in sea surface temperature in the tropical Atlantic (see Figure 2).

- Despite the apparent increases in tropical cyclone activity in Figures 2 and 3, changes in observation methods over time make it difficult to know for certain whether tropical storm activity has actually shown a long-term increase.[32]

Figure 3. North Atlantic Tropical Cyclone Activity According to the Power Dissipation Index, 1949–2011

This figure presents annual values of the Power Dissipation Index (PDI). Tropical North Atlantic sea surface temperature trends are provided for reference. Note that sea surface temperature is measured in different units, but the values have been plotted alongside the PDI to show how they compare. The lines have been smoothed using a five-year weighted average, plotted at the middle year. The most recent average (2007–2011) is plotted at 2009.

Data source: Emanuel, 2012[33]

missed before the start of aircraft and satellite observation, so scientists have used other evidence to estimate the actual number of hurricanes that might have formed in earlier years.

This indicator also looks at the Accumulated Cyclone Energy (ACE) Index and the Power Dissipation Index (PDI), which are two ways of monitoring the frequency, strength, and duration of tropical cyclones based on wind speed measurements.

Every cyclone has an ACE Index value, which is a number based on the maximum wind speed measured at six-hour intervals over the entire time that the cyclone is classified as at least a tropical storm (wind speed of at least 39 miles per hour). Therefore, a storm's ACE Index value accounts for both strength and duration. The National Oceanic and Atmospheric Administration calculates the total ACE Index value for an entire hurricane season by adding the values for all named storms, including subtropical storms, tropical storms, and hurricanes. The resulting annual total accounts for cyclone strength, duration, and frequency. For this indicator, the index has been converted to a scale where 100 equals the median value (the midpoint) over a base period from 1981 to 2010. The thresholds in Figure 2 define whether the ACE Index for a given year is close to normal, significantly above normal, or significantly below.

Like the ACE Index, the PDI is based on measurements of wind speed, but it uses a different calculation method that places more emphasis on storm intensity. This indicator shows the annual PDI value, which represents the sum of PDI values for all named storms during the year.

Indicator Notes

Over time, data collection methods have changed as technology has improved. For example, wind speed collection methods have evolved substantially over the past 60 years, while aircraft reconnaissance began in 1944 and satellite tracking around 1966. Figure 1 shows how older hurricane counts have been adjusted to attempt to account for the lack of aircraft and satellite observations. Changes in data gathering technologies could substantially influence the overall patterns in Figures 2 and 3. The effects of these changes on data consistency over the life of the indicator would benefit from additional research.

Data Sources

Hurricane counts are reported on several National Oceanic and Atmospheric Administration websites and were compiled using methods described in Knutson et al. (2010).[34] The ACE Index data (Figure 2) came from the National Oceanic and Atmospheric Administration's Climate Prediction Center, and are available online at: www.cpc.noaa.gov/products/outlooks/background_information.shtml. Values for the PDI have been calculated by Kerry Emanuel at the Massachusetts Institute of Technology. Both indices are based on wind speed measurements compiled by the National Oceanic and Atmospheric Administration.

Oceans

INDICATORS IN THIS CHAPTER

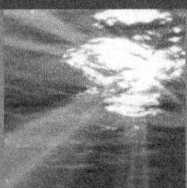 Ocean Heat

 Sea Surface Temperature

 Sea Level

The oceans and atmosphere interact constantly—both physically and chemically—exchanging energy, water, gases, and particles. This relationship influences the Earth's climate on regional and global scales. It also affects the state of the oceans.

Covering about 70 percent of the Earth's surface, the oceans store vast amounts of energy absorbed from the sun and move this energy around the globe through currents. The oceans are also a key component of the Earth's carbon cycle. Oceans store a large amount of carbon, either in dissolved form or within plants and animals (living or dead).

What is happening?

As greenhouse gases trap more energy from the sun, the oceans are absorbing more heat, resulting in an increase in sea surface temperatures and rising sea level. Although the oceans help reduce climate change by storing one-fifth to one-third of the carbon dioxide that human activities emit into the atmosphere,[1] increasing levels of dissolved carbon are changing the chemistry of seawater and making it more acidic.

Why does it matter?

Changes in ocean temperatures and currents brought about by climate change will lead to alterations in climate patterns around the world. For example, warmer waters may promote the development of stronger storms in the tropics, which can cause property damage and loss of life. Other impacts come from increased ocean acidity, which reduces the availability of some types of minerals, thus making it harder for certain organisms, such as corals and shellfish, to build their skeletons and shells. These effects, in turn, could substantially alter the biodiversity and productivity of ocean ecosystems.

Changes in ocean systems generally occur over much longer time periods than in the atmosphere, where storms can form and dissipate in a single day. Interactions between the oceans and atmosphere occur slowly over many years, and so does the movement of water within the oceans, including the mixing of deep and shallow waters. Thus, trends can persist for decades, centuries, or longer. For this reason, even if greenhouse gas emissions are stabilized tomorrow, it will take many more years—decades to centuries—for the oceans to adjust to changes in the atmosphere and the climate that have already occurred.

Ocean Acidity

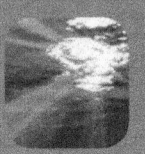

Ocean Heat

This indicator describes trends in the amount of heat stored in the world's oceans.

Background

When sunlight reaches the Earth's surface, the world's oceans absorb some of this energy and store it as heat. This heat is initially absorbed at the surface, but some of it eventually spreads to deeper waters. Currents also move this heat around the world. Water has a much higher heat capacity than air, meaning the oceans can absorb larger amounts of heat energy with only a slight increase in temperature.

The total amount of heat stored by the oceans is called "ocean heat content," and measurements of water temperature reflect the amount of heat in the water at a particular time and location. Ocean temperature plays an important role in the Earth's climate system—particularly sea surface temperature (see the Sea Surface Temperature indicator on p. 40)—because heat from ocean surface waters provides energy for storms and thereby influences weather patterns.

Higher greenhouse gas concentrations are trapping more energy from the sun, and the oceans are currently absorbing 80 to 90 percent of this extra heat— much more than the amount absorbed by the atmosphere.[2] If not for the large heat-storage capacity provided by the oceans, the atmosphere would grow warmer more rapidly.[3] Increased heat absorption also changes ocean currents because many currents are driven by differences in temperature, which causes differences in density. These currents influence climate patterns and sustain ecosystems—for example, coastal fishing grounds depend on upwelling currents to bring nutrients to the surface.

Because water expands slightly as it gets warmer, an increase in ocean heat content will also increase the volume of water in the ocean, which is one cause of the observed increases in sea level (see the Sea Level indicator on p. 42).

Figure 1. Ocean Heat Content, 1955–2011

This figure shows changes in ocean heat content between 1955 and 2011. Ocean heat content is measured in joules, a unit of energy, and compared against the 1971–2000 average, which is set at zero for reference. Choosing a different baseline period would not change the shape of the data over time. The lines were calculated independently by three agencies: the National Oceanic and Atmospheric Administration, Australia's Commonwealth Scientific and Industrial Research Organisation (CSIRO), and Japan's Agency for Marine-Earth Science and Technology (JAMSTEC).

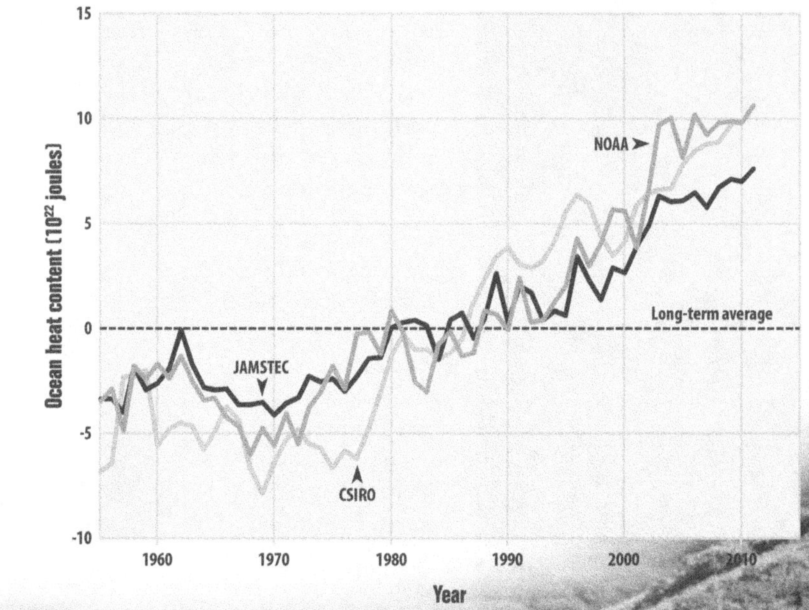

Data sources: CSIRO, 2012;[4] JAMSTEC, 2012;[5] NOAA, 2012[6]

Key Points

- In three different data analyses, the long-term trend shows that the oceans have become warmer since 1955 (see Figure 1).

- Although concentrations of greenhouse gases have risen at a relatively steady rate over the past few decades (see the Atmospheric Concentrations of Greenhouse Gases indicator on p. 16), the rate of change in ocean heat content can vary from year to year (see Figure 1). Year-to-year changes are influenced by events such as volcanic eruptions and recurring ocean-atmosphere patterns such as El Niño.

About the Indicator

This indicator shows trends in global ocean heat content from 1955 to 2011. These data are available for the top 700 meters of the ocean (nearly 2,300 feet), which accounts for just under 20 percent of the total volume of water in the world's oceans. The indicator measures ocean heat content in joules, which are a unit of energy.

The National Oceanic and Atmospheric Administration has calculated changes in ocean heat content based on measurements of ocean temperatures around the world at different depths. These measurements come from a variety of instruments deployed from ships and airplanes and, more recently, underwater robots. Thus, the data must be carefully adjusted to account for differences among measurement techniques and data collection programs. Figure 1 shows three independent interpretations of essentially the same underlying data.

Indicator Notes

Data must be carefully reconstructed and filtered for biases because of different data collection techniques and uneven sampling over time and space. Various methods of correcting the data have led to slightly different versions of the ocean heat trend line. Scientists continue to compare their results and improve their estimates over time. They also test their ocean heat estimates by looking at corresponding changes in other properties of the ocean. For example, they can check to see whether observed changes in sea level match the amount of sea level rise that would be expected based on the estimated change in ocean heat.

Data Sources

Data for this indicator were collected by the National Oceanic and Atmospheric Administration and other organizations around the world. The data were analyzed independently by researchers at the National Oceanic and Atmospheric Administration, Australia's Commonwealth Scientific and Industrial Research Organisation, and Japan's Agency for Marine-Earth Science and Technology.

Sea Surface Temperature

This indicator describes global trends in sea surface temperature.

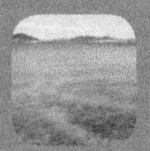

Background

Sea surface temperature—the temperature of the water at the ocean surface—is an important physical attribute of the world's oceans. The surface temperature of the world's oceans varies mainly with latitude, with the warmest waters generally near the equator and the coldest waters in the Arctic and Antarctic regions. As the oceans absorb more heat, sea surface temperatures will increase and the ocean circulation patterns that transport warm and cold water around the globe will change.

Changes in sea surface temperature can alter marine ecosystems in several ways. For example, variations in ocean temperature can affect what species of plants and animals are present in a location, alter migration and breeding patterns, threaten sensitive ocean life such as corals, and change the frequency and intensity of harmful algal blooms.[7] Over the long term, increases in sea surface temperature could also reduce the circulation patterns that bring nutrients from the deep sea to surface waters. Changes in reef habitat and nutrient supply can lead to declines in fish populations, which in turn could affect people who depend on fishing for food or jobs.[8]

Because the oceans continuously interact with the atmosphere, sea surface temperature can also have profound effects on global climate. Based on increases in sea surface temperature, the amount of atmospheric water vapor over the oceans is estimated to have increased by about 5 percent during the 20th century.[9] This water vapor feeds weather systems that produce precipitation, increasing the risk of heavy rain and snow (see the Heavy Precipitation and Tropical Cyclone Intensity indicators on pp. 30 and 34, respectively). Changes in sea surface temperature can also shift storm tracks, potentially contributing to droughts in some areas.

Figure 1. Average Global Sea Surface Temperature, 1880–2011

This graph shows how the average surface temperature of the world's oceans has changed since 1880. This graph uses the 1971 to 2000 average as a baseline for depicting change. Choosing a different baseline period would not change the shape of the data over time. The shaded band shows the range of uncertainty in the data, based on the number of measurements collected and the precision of the methods used.

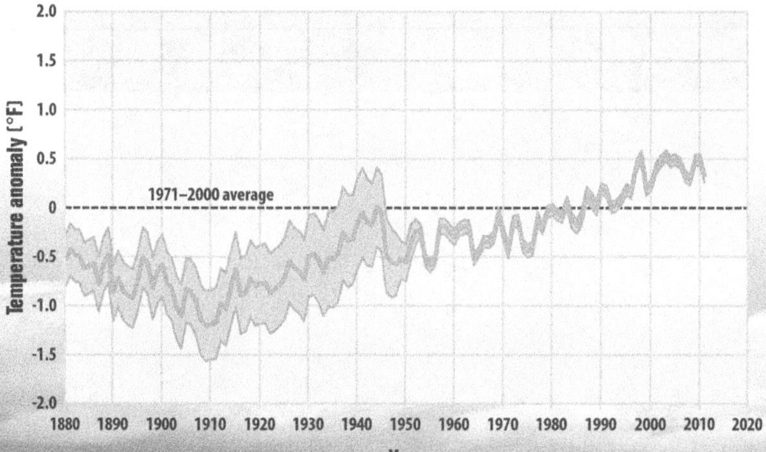

Data source: NOAA, 2012[10]

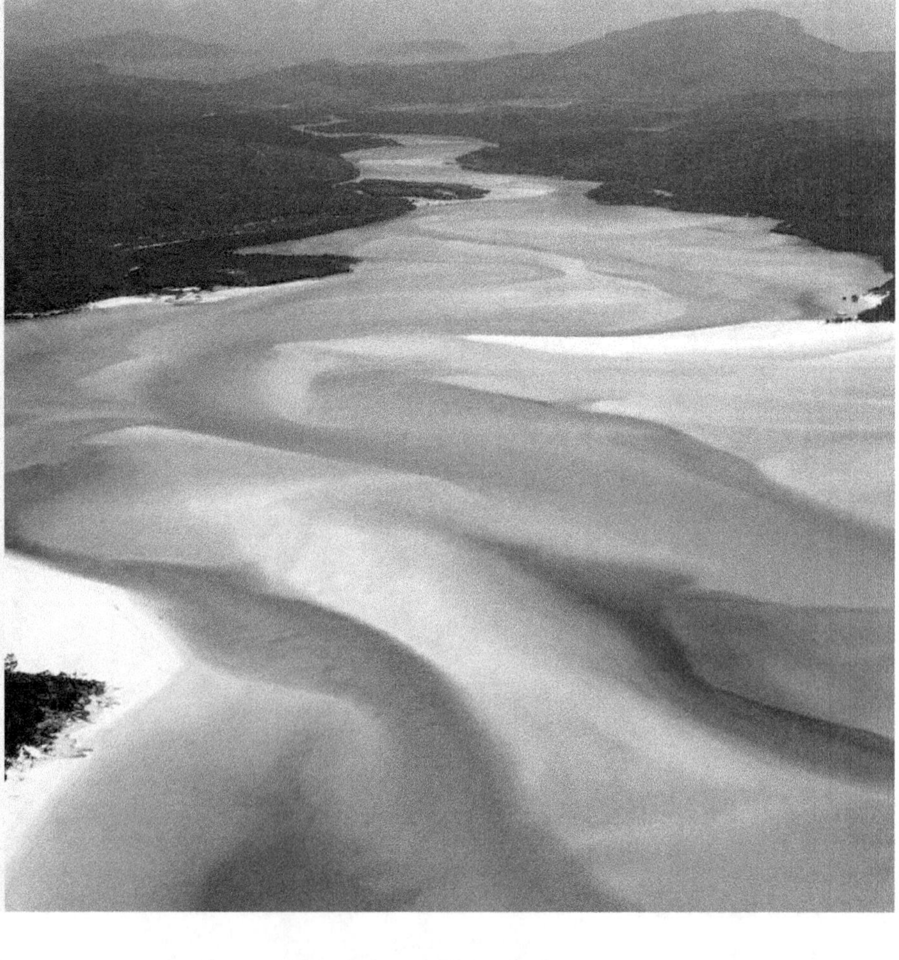

Key Points

- Sea surface temperature increased over the 20th century and continues to rise. From 1901 through 2011, temperatures rose at an average rate of 0.13°F per decade (see Figure 1).

- Sea surface temperatures have been higher during the past three decades than at any other time since reliable observations began in 1880 (see Figure 1).

- Increases in sea surface temperature have largely occurred over two key periods: between 1910 and 1940, and from 1970 to the present. Sea surface temperatures appear to have cooled between 1880 and 1910 (see Figure 1).

Example: Average Sea Surface Temperature in 2011

This map shows annual average sea surface temperatures around the world during the year 2011. It is based on a combination of direct measurements and satellite measurements.

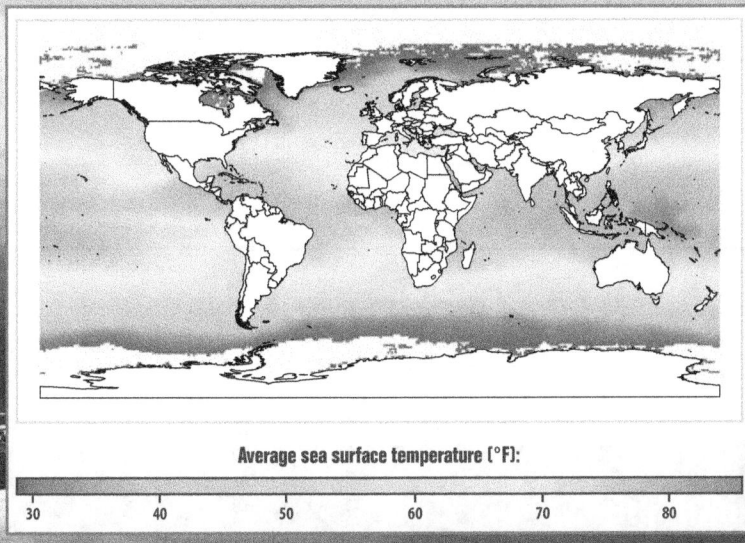

Average sea surface temperature (°F):

| 30 | 40 | 50 | 60 | 70 | 80 |

Data source: UK Met Office, 2012[ii]

About the Indicator

This indicator tracks average global sea surface temperature from 1880 through 2011 using data compiled by the National Oceanic and Atmospheric Administration. Techniques for measuring sea surface temperature have evolved since the 1800s. For instance, the earliest data were collected by inserting a thermometer into a water sample collected by lowering a bucket from a ship. Today, temperature measurements are collected more systematically from ships, as well as at stationary and drifting buoys.

The National Oceanic and Atmospheric Administration has carefully reconstructed and filtered the data for this indicator to correct for biases in the different collection techniques and to minimize the effects of sampling changes over various locations and times. The data are shown as anomalies, or differences, compared with the average sea surface temperature from 1971 to 2000.

Indicator Notes

Because this indicator tracks sea surface temperature at a global scale, the data shown in Figure 1 do not necessarily reflect local or regional trends.

Due to denser sampling and improvements in sampling design and measurement techniques, newer data are more precise than older data. The earlier trends shown by this indicator have less certainty because of lower sampling frequency and less precise sampling methods, as shown by the width of the blue shaded band in Figure 1.

Data Sources

Data for this indicator were provided by the National Oceanic and Atmospheric Administration's National Climatic Data Center and are available online at: www.ncdc.noaa.gov/ersst. These data were reconstructed from measurements of water temperature, which are available from the National Oceanic and Atmospheric Administration at: http://icoads.noaa.gov/products.html.

 # Sea Level

This indicator describes how sea level has changed over time. The indicator describes two types of sea level changes: absolute and relative.

Background

As the temperature of the Earth changes, so does sea level. Temperature and sea level are linked for two main reasons:

1. Changes in the volume of water and ice on land (namely glaciers and ice sheets) can increase or decrease the volume of water in the ocean (see the Glaciers indicator on p. 50).

2. As water warms, it expands slightly—an effect that is cumulative over the entire depth of the oceans (see the Ocean Heat indicator on p. 38).

Changing sea levels can affect human activities in coastal areas. For example, rising sea levels can lead to increased coastal flooding and erosion, which is a particular concern in low-lying areas. Higher sea level also makes coastal infrastructure more vulnerable to damage from storms. Sea level rise can alter eco-systems, transforming marshes and other wetlands into open water and freshwater systems into salt water systems.

The sea level changes that affect coastal systems involve more than just expand-ing oceans, however, because the Earth's continents can also rise and fall relative to the oceans. Land can rise through pro-cesses such as sediment accumulation (the process that built the Mississippi Delta) and geological uplift (for example, as gla-ciers melt and the land below is no longer weighed down by heavy ice). In other areas, land can sink because of erosion, sediment compaction, natural subsidence (sinking due to geologic changes), or engi-neering projects that prevent rivers from naturally depositing sediments along their banks. Changes in ocean currents such as the Gulf Stream can also affect sea levels by pushing more water against some coastlines and pulling it away from others, raising or lowering sea levels accordingly.

Scientists account for these types of changes by measuring sea level change in two different ways. *Relative* sea level change is how the height of the ocean rises or falls relative to the land at a particular location. In contrast, *absolute* sea level change refers to the height of the ocean surface above the center of the earth, without regard to whether nearby land is rising or falling.

Figure 1. Global Average Absolute Sea Level Change, 1880–2011

This graph shows cumulative changes in sea level for the world's oceans since 1880, based on a combination of long-term tide gauge measurements and recent satellite measurements. This figure shows average absolute sea level change, which refers to the height of the ocean surface, regardless of whether nearby land is rising or falling. Satellite data are based solely on measured sea level, while the long-term tide gauge data include a small correction factor because the size and shape of the oceans are changing slowly over time. (On average, the ocean floor has been gradually sinking since the last Ice Age peak, 20,000 years ago.) The shaded band shows the likely range of values, based on the number of measurements collected and the precision of the methods used.

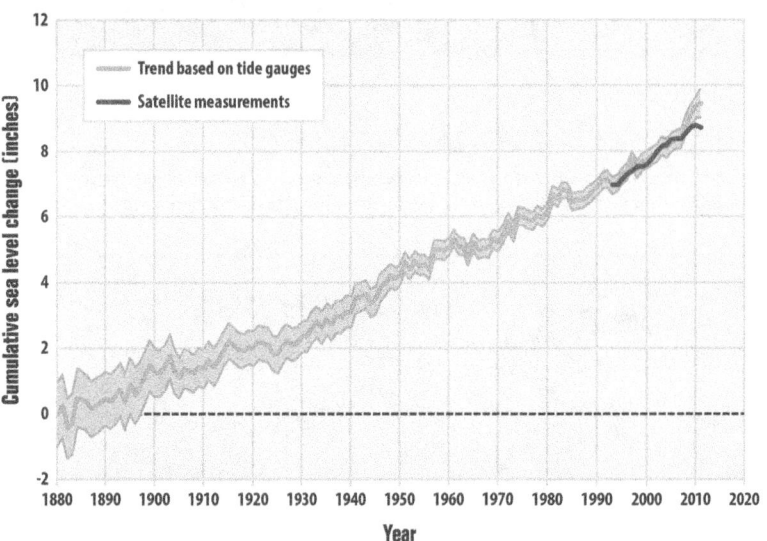

Data sources: CSIRO, 2012;[12] NOAA, 2012[13]

Key Points

*   After a period of approximately 2,000 years of little change (not shown here), global aver-age sea level rose throughout the 20th century, and the rate of change has accelerated in re-cent years.[14] When averaged over all the world's oceans, absolute sea level increased at an average rate of 0.07 inches per year from 1880 to 2011 (see Figure 1). From 1993 to 2011, however, average sea level rose at a rate of 0.11 to 0.13 inches per year—roughly twice as fast as the long-term trend.

*   Relative sea level rose along much of the U.S. coastline between 1960 and 2011, particu-larly the Mid-Atlantic Coast and parts of the Gulf Coast, where some stations registered increases of more than 8 inches (see Figure 2). Meanwhile, relative sea level fell at some locations in Alaska and the Pacific Northwest. At those sites, even though absolute sea level has risen, land elevation has risen more rapidly.

*   While absolute sea level has increased steadily overall, particularly in recent decades, re-gional trends vary, and absolute sea level has decreased in some places.[15] Relative sea level also has not risen uniformly because of regional and local changes in land movement and long-term changes in coastal circulation patterns.

Figure 2. Relative Sea Level Change Along U.S. Coasts, 1960–2011

This map shows cumulative changes in relative sea level from 1960 to 2011 at tide gauge stations along U.S. coasts. Relative sea level reflects changes in sea level as well as land elevation.

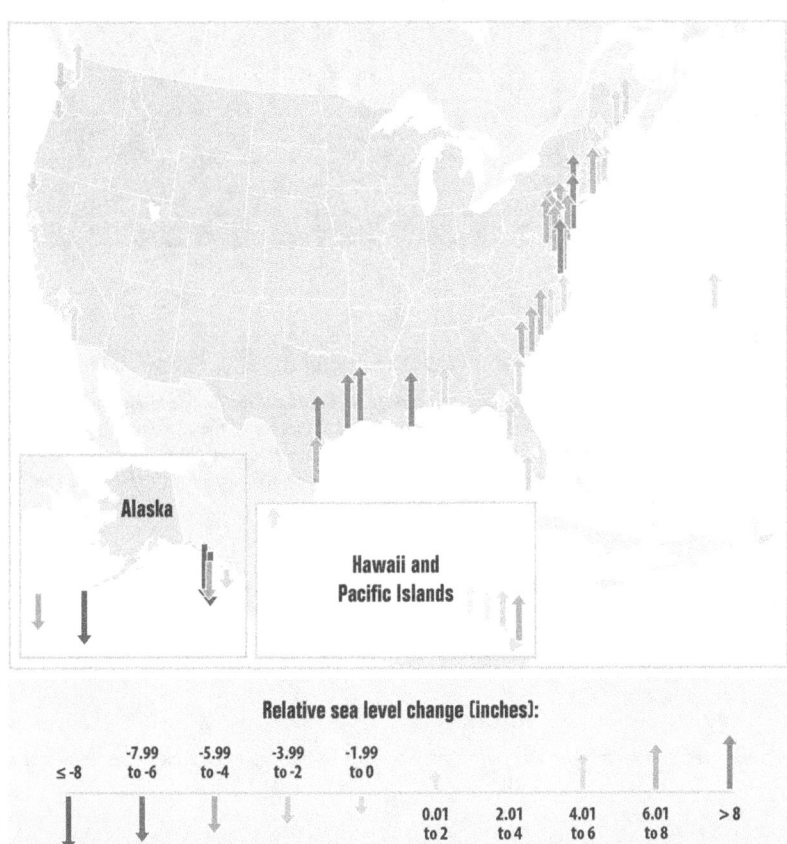

Data source: NOAA, 2012[16]

About the Indicator

This indicator presents trends in sea level based on measurements from tide gauges and from satellites that orbit the Earth. Tide gauges measure relative sea level change at points along the coast, while satellite instruments measure absolute sea level change over nearly the entire ocean surface. Many tide gauges have collected data for more than 100 years, while satellites have collected data since the early 1990s.

Figure 1 shows annual absolute sea level change averaged over the entire Earth's ocean surface. The long-term trend is based on tide gauge data that have been adjusted to show absolute global trends through calibration with recent satellite data. Figure 2 shows trends at a more local scale, highlighting the 1960 to 2011 change in relative sea level at 68 tide gauges along the Atlantic, Pacific, and Gulf coasts of the United States.

Indicator Notes

Relative sea level trends represent a combination of absolute sea level change and any local land movement. Tide gauge measurements such as those in Figure 2 generally cannot distinguish between these two different influences without an accurate measurement of vertical land motion nearby.

Some changes in relative and absolute sea level can be due to multi-year cycles such as El Niño and La Niña, which affect coastal ocean temperatures, salt content, wind patterns, atmospheric pressure (and thus storm tracks), and currents. Obtaining a reliable trend can require many years of data, which is why the satellite record in Figure 1 has been supplemented with a longer-term reconstruction based on tide gauge measurements.

Data Sources

Absolute sea level trends were provided by Australia's Commonwealth Scientific and Industrial Research Organisation and the National Oceanic and Atmospheric Administration. These data are based on measurements collected by satellites and tide gauges. Relative sea level data are available from the National Oceanic and Atmospheric Administration, which publishes an interactive online map (http://tidesandcurrents.noaa.gov/sltrends/sltrends.shtml) with links to detailed data for each tide gauge.

Ocean Acidity

Background

The ocean plays an important role in regulating the amount of carbon dioxide in the atmosphere. As atmospheric concentrations of carbon dioxide rise (see the Atmospheric Concentrations of Greenhouse Gases indicator on p. 16), the ocean absorbs more carbon dioxide. Because of the slow mixing time between surface waters and deeper waters, it can take hundreds to thousands of years to establish this balance. Over the past 250 years, oceans have absorbed approximately 40 percent of the carbon dioxide produced by human activities.[17]

Although the ocean's ability to take up carbon dioxide prevents atmospheric levels from climbing even higher, rising levels of carbon dioxide dissolved in the ocean can have a negative effect on marine life. Carbon dioxide reacts with sea water to produce carbonic acid. The resulting increase in acidity (measured by lower pH values) reduces the availability of minerals such as aragonite, which is a form of calcium carbonate that corals, some types of plankton, and other creatures rely on to produce their hard skeletons and shells. Declining pH and reduced availability of minerals can make it more difficult for these animals to thrive. This can lead to broader changes in the overall structure of ocean and coastal ecosystems, and can ultimately affect fish populations and the people who depend on them.[18]

While changes in ocean pH and mineral availability caused by the uptake of atmospheric carbon dioxide generally occur over many decades, these properties can fluctuate over shorter periods, especially in coastal and surface waters. For example, increased photosynthesis and respiration during the day and during the summer leads to natural fluctuations in pH. Acidity also varies with water temperature.

About the Indicator

This indicator describes trends in pH and related properties of ocean water, based on a combination of direct observations, calculations, and modeling.

Figure 1 shows pH values and levels of dissolved carbon dioxide at three locations that have collected measurements consistently over the last few decades. These data have been either measured directly or calculated from related measurements such as dissolved

Key Points

- Measurements made over the last few decades have demonstrated that ocean carbon dioxide levels have risen in response to increased carbon dioxide in the atmosphere, leading to an increase in acidity (that is, a decrease in pH) (see Figure 1).

- Historical modeling suggests that since the 1880s, increased carbon dioxide has led to lower aragonite saturation levels (less availability of minerals) in the oceans around the world (see Figure 2).

- The largest decreases in aragonite saturation have occurred in tropical waters (see Figure 2). However, decreases in cold areas may be of greater concern because colder waters typically have lower aragonite levels to begin with.[19]

Figure 1. Ocean Carbon Dioxide Levels and Acidity, 1983–2011

This figure shows the relationship between changes in ocean carbon dioxide levels (measured in the left column as a partial pressure—a common way of measuring the amount of a gas) and acidity (measured as pH in the right column). The data come from two observation stations in the North Atlantic Ocean (Canary Islands and Bermuda) and one in the Pacific (Hawaii). The up-and-down pattern shows the influence of seasonal variations.

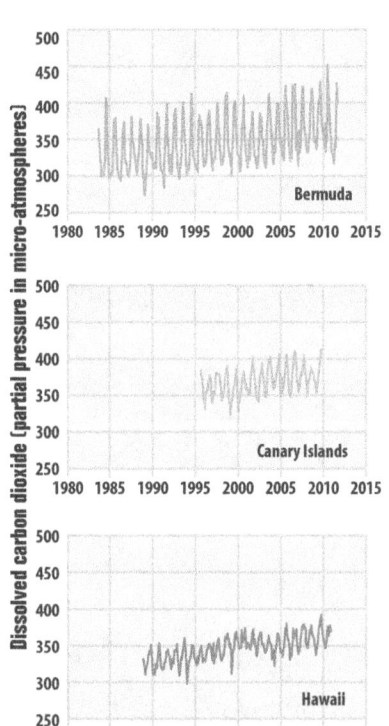
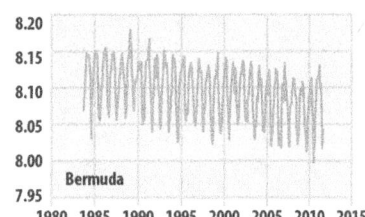
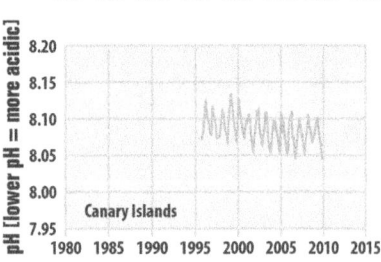
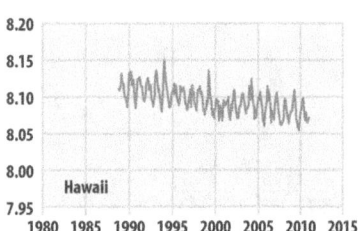

Data sources: Bates et al., 2012;[20] González-Dávila, 2012;[21] University of Hawaii, 2012[22]

(Continued on page 45)

pH Scale

Acidity is commonly measured using the pH scale. Pure water has a pH of about 7, which is considered neutral. A substance with a pH less than 7 is considered to be acidic, while a substance with a pH greater than 7 is considered to be basic or alkaline. The lower the pH, the more acidic the substance. Like the well-known Richter scale for measuring earthquakes, the pH scale is based on powers of 10, which means a substance with a pH of 3 is 10 times more acidic than a substance with a pH of 4. For more information about pH, visit: www.epa.gov/acidrain/measure/ph.html.

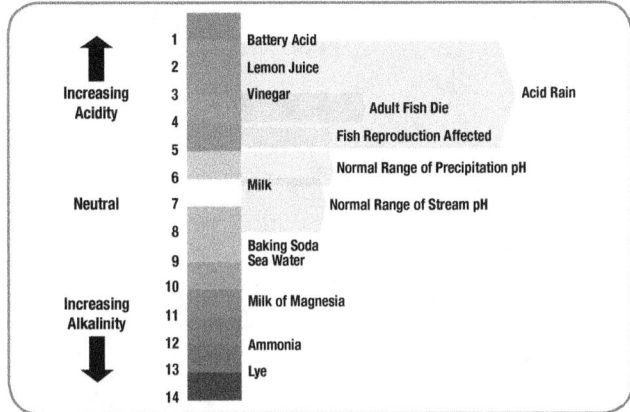

Source: Environment Canada, 2008[23]

Figure 2. Changes in Aragonite Saturation of the World's Oceans, 1880–2012

This map shows changes in the amount of aragonite dissolved in ocean surface waters between the 1880s and the most recent decade (2003–2012). Aragonite is a form of calcium carbonate that many marine animals use to build their skeletons and shells. Aragonite saturation is a ratio that compares the amount of aragonite that is actually present with the total amount of aragonite that the water could hold if it were completely saturated. The more negative the change in aragonite saturation, the larger the decrease in aragonite available in the water, and the harder it is for marine creatures to produce their skeletons and shells.

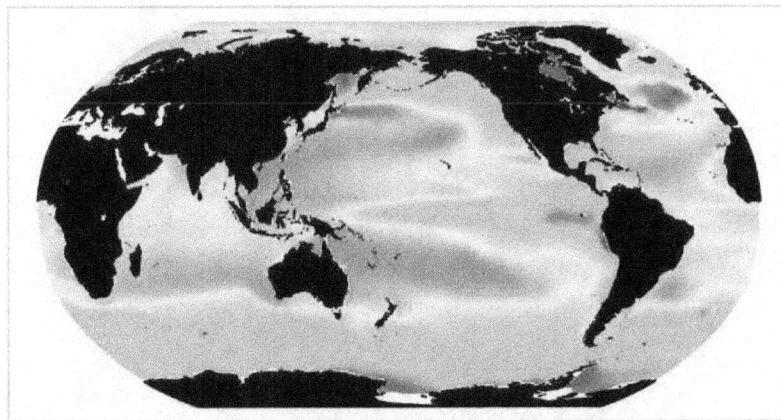

Change in aragonite saturation at the ocean surface (Ω_{ar}):

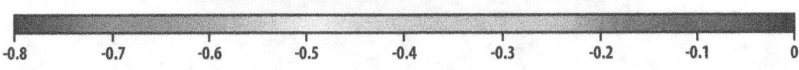

Data source: Feely et al., 2009[24]

inorganic carbon and alkalinity. Data come from two stations in the Atlantic Ocean (Bermuda and the Canary Islands) and one in the Pacific (Hawaii).

The global map in Figure 2 shows changes over time in the amount of aragonite dissolved in ocean water, which is called aragonite saturation. This map was created by comparing average conditions during the 1880s with average conditions during the most recent 10 years (2003–2012). Aragonite saturation has only been measured at selected locations during the last few decades, but it can be calculated reliably for different times and locations based on the relationships scientists have observed among aragonite saturation, pH, dissolved carbon, water temperature, concentrations of carbon dioxide in the atmosphere, and other factors that can be measured. Thus, while Figure 2 was created using a computer model, it is indirectly based on actual measurements.

Indicator Notes

This indicator focuses on surface waters, which absorb carbon dioxide from the atmosphere within a few months.[25] It can take much longer for changes in pH and mineral saturation to spread to deeper waters, so the full effect of increased atmospheric carbon dioxide concentrations on ocean acidity may not be seen for many decades, if not centuries. Studies suggest that the impacts of ocean acidification may be greater at depth, because the availability of minerals like aragonite is naturally lower in deeper waters.[26]

Ocean chemistry is not uniform around the world, so local conditions can cause pH or aragonite saturation measurements to differ from the global average. For example, carbon dioxide dissolves more readily in cold water than in warm water, so colder regions could experience greater impacts from acidity than warmer regions.

Data Sources

Data for Figure 1 came from three studies: the Bermuda Atlantic Time-Series Study, the European Station for Time-Series in the Ocean (Canary Islands), and the Hawaii Ocean Time-Series. Bermuda data were analyzed by Bates et al. (2012)[27] and are available at: http://bats.bios.edu. Canary Islands data were analyzed by Gonzalez-Davila et al. (2010)[28] and are available at: www.eurosites.info/estoc/data.php. Hawaii data were analyzed by Dore et al. (2009)[29] and are available at: http://hahana.soest. hawaii.edu/hot/products/products.html.

The map in Figure 2 was created by the National Oceanic and Atmospheric Administration and the Woods Hole Oceanographic Institution using Community Earth System Model data. Related information can be found at: http://sos.noaa.gov/Datasets/list. php?category=Ocean.

Snow and Ice

INDICATORS IN THIS CHAPTER

 Arctic Sea Ice

Glaciers

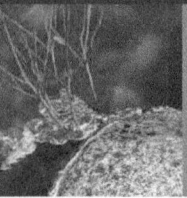 Lake Ice

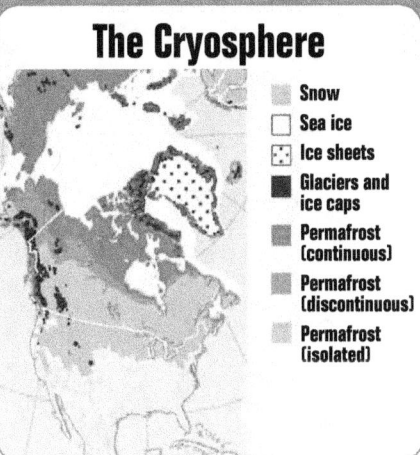

The Cryosphere

- Snow
- Sea ice
- Ice sheets
- Glaciers and ice caps
- Permafrost (continuous)
- Permafrost (discontinuous)
- Permafrost (isolated)

Source: UNEP, 2007[1]

The Earth's surface contains many forms of snow and ice, including sea ice, lake and river ice, snow cover, glaciers, ice caps and sheets, and frozen ground. Together, these features are sometimes referred to as the "cryosphere," a term for all parts of the Earth where water exists in solid form.

Snow and ice are an important part of the global climate system. Because snow and ice are highly reflective, much of the sunlight that hits these surfaces is reflected back into space instead of warming the Earth. The presence or absence of snow and ice affects heating and cooling over the Earth's surface, influencing the planet's energy balance.

Climate change can dramatically alter the Earth's snow- and ice-covered areas. Unlike other substances found on the Earth, snow and ice exist at temperatures close to their melting point and can thus change between solid and liquid states in response to relatively minor changes in temperature. As a result, prolonged warming or cooling trends can result in significant changes across the landscape as snow and ice masses shrink or grow over time.

What is happening?

Some regions that usually receive snow are receiving less snowfall and do not have as much snow on the ground. Glaciers in the United States and around the world have generally shrunk, and the rate at which they are melting appears to have accelerated over the last decade. Additionally, the amount of ice in the Arctic Ocean has decreased, and many lakes are freezing later in the fall and melting earlier in the spring.

Why does it matter?

Reduced snowfall and less snow cover on the ground could diminish the beneficial insulating effects of snow for vegetation and wildlife, while also affecting water supplies, transportation, cultural practices, travel, and recreation for millions of people. For communities in Arctic regions, reduced sea ice could increase coastal erosion and exposure to storms, threatening homes and property, while thawing ground could damage roads and buildings and accelerate erosion.

Such changing climate conditions can have worldwide implications because snow and ice influence air temperatures, sea level, ocean currents, and storm patterns. For example, melting ice sheets on Greenland and Antarctica add fresh water to the ocean, increasing sea level and possibly changing ocean circulation that is driven by differences in temperature and salinity. Because of their light color, snow and ice also reflect more sunlight than open water or bare ground, so a reduction in snow cover and ice causes the Earth's surface to absorb more energy from the sun.

Thawing of frozen ground and reduced sea ice in the Arctic could affect biodiversity on local and global scales, leading to harmful effects not only on polar bears and seals, but also on migratory species that breed or feed in these areas. These changes could affect people by compromising their livelihoods and traditional means of gathering food, particularly Arctic indigenous populations. Conversely, reduced snow and ice could present commercial opportunities for others, including ice-free shipping lanes and increased access to natural resources.

 Snowfall

 Snow Cover

 Snowpack

Arctic Sea Ice

This indicator tracks the extent and age of sea ice in the Arctic Ocean.

Background

Sea ice is an integral part of the Arctic Ocean. During the dark winter months, sea ice essentially covers the entire Arctic Ocean. In summer, some of this ice melts because of warmer temperatures and long hours of sunlight. Sea ice typically reaches its minimum thickness and extent in mid-September, when the area covered by ice is roughly half the size of the winter maximum. The ice then begins expanding again.

The extent of area covered by Arctic sea ice is an important indicator of changes in global climate because warmer air and water temperatures are reducing the amount of sea ice present. Because sea ice is more reflective than liquid water, it plays a significant role in the Earth's energy balance and keeping polar regions cool. (For more information on the effects of surface color on reflecting sunlight, see the Snow Cover indicator on p. 56.) Sea ice also keeps the air cool by forming a barrier between the air above and the warmer water below. As the amount of sea ice decreases, the Arctic region's ability to stabilize the Earth's climate is reduced, potentially leading to a "feedback loop" of more absorption of solar energy, higher air temperatures, and even greater loss of sea ice.

The age of sea ice is also an important indicator of Arctic conditions because older ice is generally thicker and stronger than younger ice. A loss of older ice suggests that the Arctic is losing ice faster than it is accumulating it.

Changes in sea ice can directly affect the health of Arctic ecosystems. Mammals such as polar bears and walruses rely on the presence of sea ice to preserve their hunting, breeding, and migrating habits. These animals face the threat of declining birth rates and restricted access to food sources because of reduced sea ice coverage and thickness. Impacts on Arctic wildlife, as well as the loss of ice itself, are already restricting the traditional subsistence hunting lifestyle of indigenous Arctic populations such as the Yup'ik, Iñupiat, and Inuit.

While diminished sea ice can have negative ecological effects, it can also present commercial opportunities. For instance, reduced sea ice opens shipping lanes and increases access to natural resources in the Arctic region.

Dwindling Arctic Sea Ice

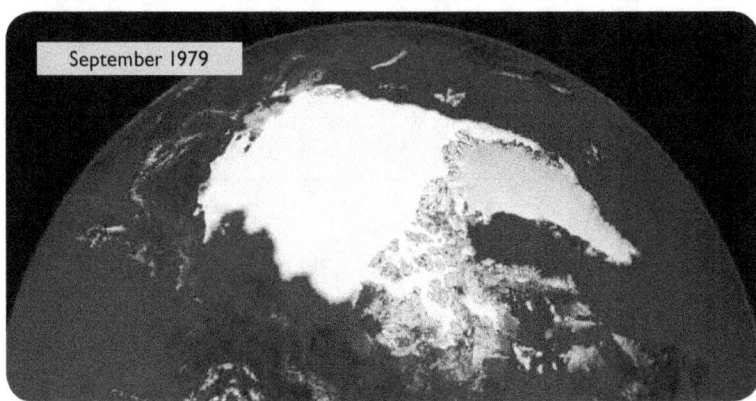

September 1979

September 2012

Source: NASA, 2012[2]

Key Points

- September 2012 had the lowest sea ice extent on record, 49 percent below the 1979–2000 average for that month.

- The September 2012 record low sea ice extent was 1.3 million square miles (an area five times the size of Texas) less than the historical 1979–2000 average (see Figure 1).

- Although the annual minimum of sea ice extent typically occurs in September, all months have shown a decreasing trend in sea ice extent over the past several decades. The largest decreases have occurred in the summer and fall.[3, 4]

- Evidence of the age of Arctic sea ice suggests an overall loss of multi-year ice. The proportion of sea ice five years or older has declined dramatically over the recorded time period, from more than 30 percent of September ice in the 1980s to 4 percent in 2012. A growing percentage of Arctic sea ice is only one or two years old. This thinning of Arctic ice makes it more vulnerable to further melting.

Figure 1. September Monthly Average Arctic Sea Ice Extent, 1979–2012

This figure shows Arctic sea ice extent from 1979 through 2012 using data from September of each year, which is when the minimum extent typically occurs.

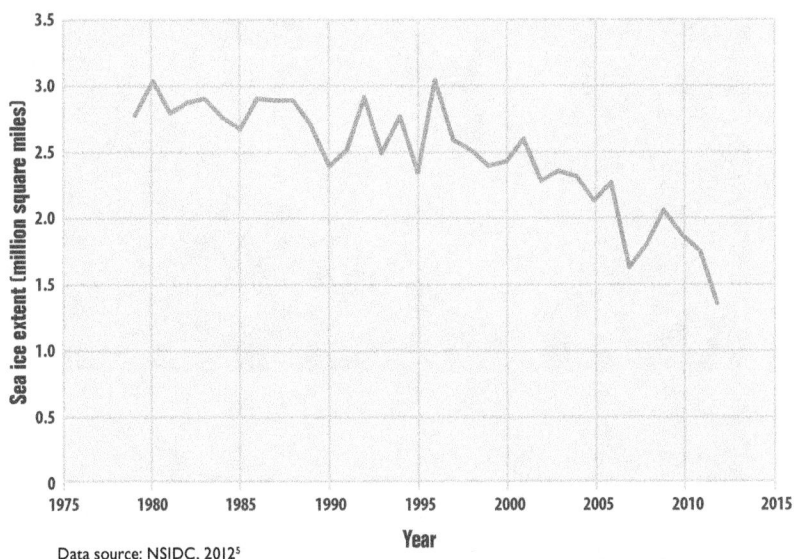

Data source: NSIDC, 2012[5]

Figure 2. Age of Arctic Sea Ice at Minimum September Week, 1983–2012*

This figure shows the distribution of Arctic sea ice extent by age group during the peak melting week in September of each year.

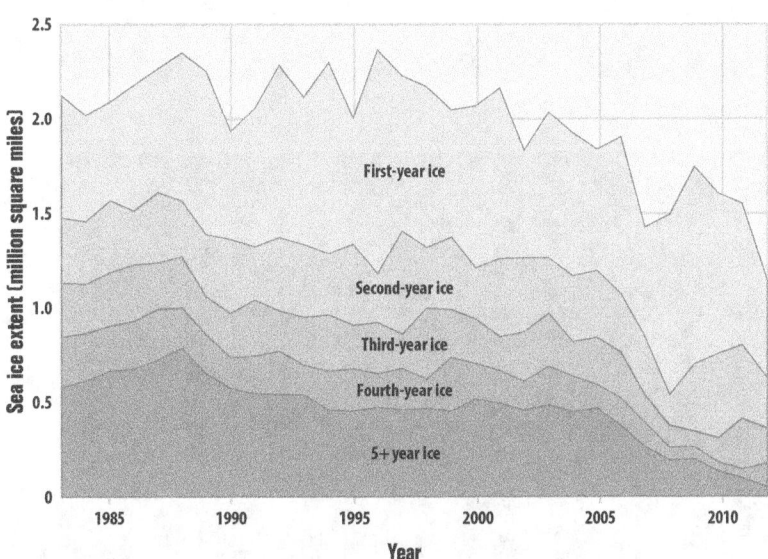

* The total extent in Figure 2 differs from Figure 1 because Figure 1 shows a monthly average, while Figure 2 shows conditions during a single week.

Data source: NSIDC, 2012[6]

About the Indicator

Figure 1 presents trends in Arctic sea ice extent from 1979, when extensive measurements started, to 2012. Sea ice extent is defined as the area of ocean where at least 15 percent of the surface is frozen. This threshold was chosen because scientists have found that it gives the best approximation of the edge of the ice. Data are collected throughout the year, but for comparison, this indicator focuses on the average sea ice extent in September of each year. This is because September is typically when the sea ice extent reaches its annual minimum after melting during the spring and summer. Data for this indicator were gathered by the National Snow and Ice Data Center using satellite imaging technology.

Figure 2 examines the age of the ice that is present in the Arctic during the week in September with the smallest extent of ice. By combining daily satellite images, wind measurements, and data from surface buoys that move with the ice, scientists can track specific parcels of ice as they move over time. This tracking enables them to calculate the age of the ice in different parts of the Arctic. Although satellites started collecting data in 1979, Figure 2 only shows trends back to 1983 because it is not possible to know the full age distribution until the ice has been tracked for at least five years.

Indicator Notes

Increasing temperatures associated with climate change are not the only factor contributing to reductions in sea ice. Other conditions that may be affected by climate change, such as fluctuations in oceanic and atmospheric circulation and typical annual and decadal variability, also affect the extent of sea ice. Determining the age of ice is an imperfect science, as there are cases where a small amount of older ice might exist within an area classified as younger, or vice-versa.

Data Sources

The data for this indicator were provided by the National Snow and Ice Data Center. Data for Figure 1 are also available online at: http://nsidc.org/data/seaice_index/archives/index.html, while Figure 2 is based on an analysis by the University of Colorado and a graph published at: http://nsidc.org/arctic-seaicenews/2012/10/poles-apart-a-record-breaking-summer-and-winter. The National Snow and Ice Data Center produces a variety of reports and a seasonal newsletter analyzing Arctic sea ice data.

Glaciers

This indicator examines the balance between snow accumulation and melting in glaciers, and it describes how glaciers around the world have changed over time.

Background

A glacier is a large mass of snow and ice that has accumulated over many years and is present year-round. In the United States, glaciers can be found in the Rocky Mountains, the Sierra Nevada, the Cascades, and throughout Alaska. A glacier flows naturally like a river, only much more slowly. At higher elevations, glaciers accumulate snow, which eventually becomes compressed into ice. At lower elevations, the "river" of ice naturally loses mass because of melting and ice breaking off and floating away (iceberg calving) if the glacier ends in a lake or the ocean. When melting and calving are exactly balanced by new snow accumulation, a glacier is in equilibrium and its mass will neither increase nor decrease.

In many areas, glaciers provide communities and ecosystems with a reliable source of streamflow and drinking water, particularly in times of extended drought and late in the summer, when seasonal snowpack has melted away. Freshwater runoff from glaciers also influences ocean ecosystems. Glaciers are important as an indicator of climate change because physical changes in glaciers—whether they are growing or shrinking, advancing or receding—provide visible evidence of changes in temperature and precipitation. If glaciers lose more ice than they can accumulate through new snowfall, they ultimately add more water to the oceans, leading to a rise in sea level (see the Sea Level indicator on p. 42). The same kinds of changes occur on a much larger scale within the giant ice sheets that cover Greenland and Antarctica, potentially leading to even bigger implications for sea level. Small glaciers tend to respond more quickly to climate change than the giant ice sheets, however, and they have added more water to the oceans than the ice sheets have in recent decades.[7]

About the Indicator

This indicator is based on long-term monitoring data collected at selected glaciers around the world. Scientists collect detailed measurements to determine glacier mass balance, which is the net gain or loss of snow and ice over the course of the year. A negative mass balance indicates that a glacier has lost ice or snow. The cumulative mass balance over time reveals long-term trends. For example, if cumulative mass balance becomes more negative over time, it means glaciers are losing mass more quickly than they can accumulate new snow.

(Continued on page 51)

Photographs of McCall Glacier, Alaska, 1958 and 2003

1958

2003

Sources: Post, 1958;[8] Nolan, 2003[9]

Key Points

- On average, glaciers worldwide have been losing mass since at least the 1970s (see Figure 1), which in turn has contributed to observed changes in sea level (see the Sea Level indicator on p. 42). Measurements from a smaller number of glaciers suggest that they have been shrinking since the 1940s. The rate at which glaciers are losing mass appears to have accelerated over roughly the last decade.

- All three U.S. benchmark glaciers have shown an overall decline in mass balance since the 1950s and 1960s and an accelerated rate of decline in recent years (see Figure 2). Year-to-year trends vary, with some glaciers gaining mass in certain years (for example, Wolverine Glacier during the 1980s). However, most of the measurements indicate a loss of glacier mass over time.

- Trends for the three benchmark glaciers are consistent with the retreat of glaciers observed throughout the western United States, Alaska, and other parts of the world.[10] Observations of glaciers losing mass are also consistent with warming trends in U.S. and global temperatures during this time period (see the U.S. and Global Temperature indicator on p. 24).

Figure 1. Average Cumulative Mass Balance of "Reference" Glaciers Worldwide, 1945–2010

This figure shows the cumulative change in mass balance of a set of "reference" glaciers worldwide beginning in 1945. The line on the graph represents the average of all the glaciers that were measured. Negative values in later years indicate a net loss of ice and snow compared with the base year of 1945. For consistency, measurements are in meters of water equivalent, which represent changes in the average thickness of a glacier. The small chart below shows how many glaciers were measured in each year. Some glacier measurements have not yet been finalized for 2010, hence the smaller number of sites.

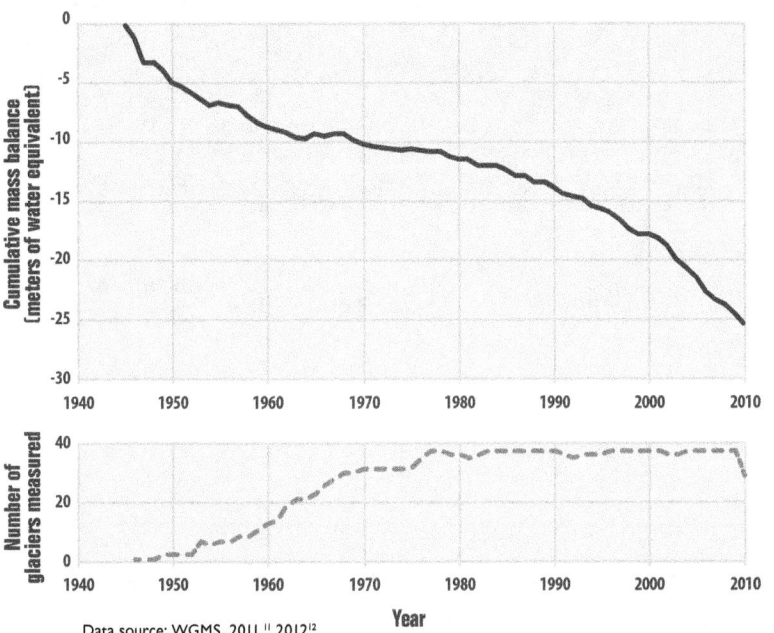

Data source: WGMS, 2011,[11] 2012[12]

Figure 2. Cumulative Mass Balance of Three U.S. Glaciers, 1958–2010

This figure shows the cumulative mass balance of the three U.S. Geological Survey "benchmark" glaciers since measurements began in the 1950s or 1960s. For each glacier, the mass balance is set at zero for the base year of 1965. Negative values in later years indicate a net loss of ice and snow compared with the base year. For consistency, measurements are in meters of water equivalent, which represent changes in the average thickness of a glacier. The dashed line in the lower right corner represents a preliminary number for 2010.

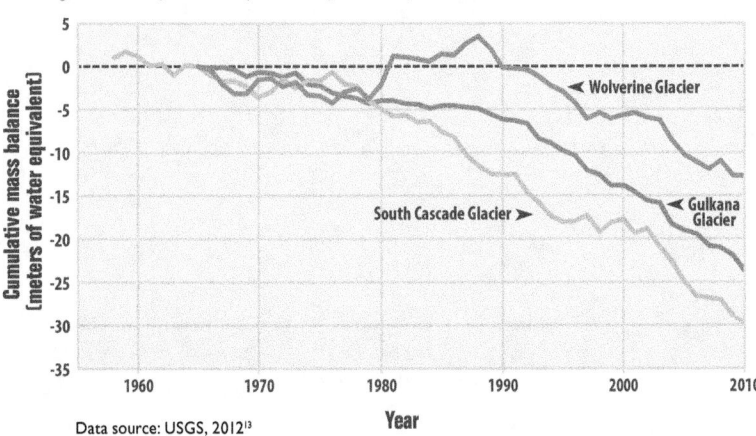

Data source: USGS, 2012[13]

Glaciers Shown in Figure 2

AK Gulkana Glacier

Wolverine Glacier

South Cascade Glacier
WA

Figure 1 shows trends in mass balance for a set of 37 reference glaciers around the world that have been measured consistently since the 1970s, including a few that have been measured since the 1940s. Data from these reference glaciers have been averaged together to depict changes over time. Figure 2 shows trends for three "benchmark" glaciers: South Cascade Glacier in Washington state, Wolverine Glacier near Alaska's southern coast, and Gulkana Glacier in Alaska's interior. These three glaciers were chosen because they have been studied extensively by the U.S. Geological Survey for many years and because they are thought to be representative of other glaciers nearby.

This indicator describes the change in glacier mass balance, which is measured as the average change in thickness across the surface of a glacier. The change in ice or snow has been converted to the equivalent amount of liquid water.

Indicator Notes

The relationship between climate change and glacier mass balance is complex, and the observed changes at specific reference or benchmark glaciers might reflect a combination of global and local climate variations. Slightly different measurement and analysis methods have been used at different glaciers, but overall trends appear to be similar.

Long-term measurements are available for only a relatively small percentage of the world's glaciers. This indicator does not include the Greenland and Antarctic ice sheets, although nearly two decades of satellite data suggest that these ice sheets are also experiencing a net loss of ice.[14] Continued satellite data collection will allow scientists to evaluate long-term trends in the future.

Data Sources

The World Glacier Monitoring Service compiled data for Figure 1, based on measurements collected by a variety of organizations around the world. The U.S. Geological Survey Benchmark Glacier Program provided the data for Figure 2. These data, as well as periodic reports and measurements of the benchmark glaciers, are available on the program's website at: http://ak.water.usgs.gov/glaciology.

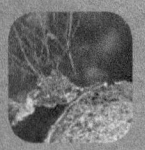

Lake Ice

Background

The formation of ice cover on lakes in the winter and its disappearance the following spring depends on climate factors such as air temperature, cloud cover, and wind. Conditions such as heavy rains or snowmelt in locations upstream or elsewhere in the watershed also affect lake ice duration. Thus, ice formation and breakup dates are key indicators of climate change. If lakes remain frozen for longer periods, it can signify that the climate is cooling. Conversely, shorter periods of ice cover suggest a warming climate.

Changes in ice cover can affect the physical, chemical, and biological characteristics of a body of water. For example, ice influences heat and moisture transfers between a lake and the atmosphere. Reduced ice cover leads to increased evaporation and lower water levels, as well as an increase in water temperature and sunlight penetration. These changes, in turn, can affect plant and animal life cycles and the availability of suitable habitat. Additionally, ice cover affects the amount of heat that is reflected from the Earth's surface. Exposed water will absorb and retain heat, whereas an ice- and snow-covered lake will reflect the sun's energy rather than absorb it. (For more information on ice and snow reflecting sunlight, see the Snow Cover indicator on p. 56.)

The timing and duration of ice cover on lakes and other bodies of water can also affect society—particularly shipping and transportation, hydroelectric power generation, and fishing. The impacts can be either positive or negative. For example, reduced ice cover on a large lake could extend the open-water shipping season but require vessels to reduce their cargo capacity, as increased evaporation leads to lower water levels.

About the Indicator

This indicator analyzes the dates at which lakes freeze and thaw. Freeze dates are when a continuous and immobile ice cover forms over a body of water. Thaw dates are when the ice

(Continued on page 53)

Key Points

- The time that lakes stay frozen has generally decreased since the mid-1800s. For most of the lakes in this indicator, the duration of ice cover has decreased at an average rate of one to two days per decade (see Figure 1).

- The lakes covered by this indicator are generally freezing later than they did in the past. Freeze dates have grown later at a rate of roughly half a day to one day per decade (see Figure 2).

- Thaw dates for most of these lakes show a general trend toward earlier ice breakup in the spring (see Figure 3).

- The changes in freeze and thaw dates shown here are consistent with other studies. For example, a broad study of lakes and rivers throughout the Northern Hemisphere found that since the mid-1800s, freeze dates have occurred later, at an average rate of 5.8 days per 100 years, and thaw dates have occurred earlier, at an average rate of 6.5 days per 100 years.[15]

Figure 1. Duration of Ice Cover for Selected U.S. Lakes, 1850–2010

This figure displays the duration (in days) of ice cover for eight U.S. lakes. The data are available from approximately 1850 to 2010, depending on the lake, and have been smoothed using a nine-year moving average.

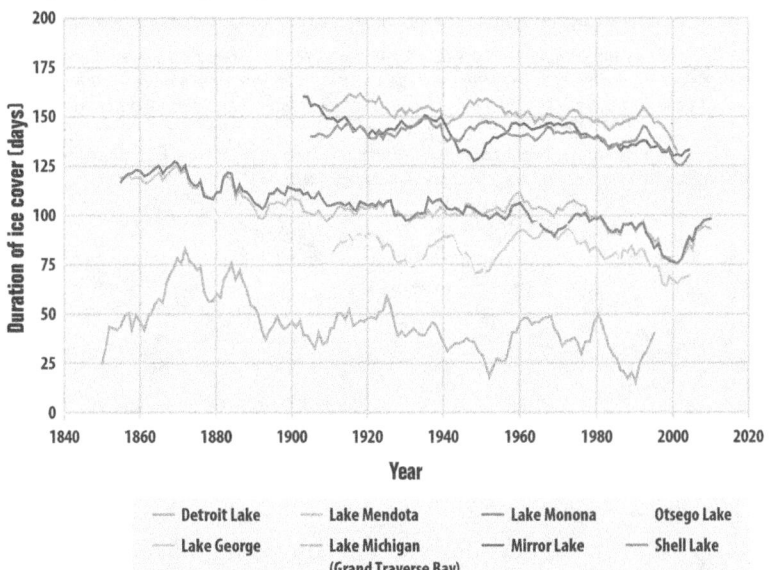

Data source: NSIDC, 2011[16]

Figure 2. Date of First Freeze for Selected U.S. Lakes, 1850–2010

This figure shows the "ice-on" date, or date of first freeze, for eight U.S. lakes. The data are available from approximately 1850 to 2010, depending on the lake, and have been smoothed using a nine-year moving average.

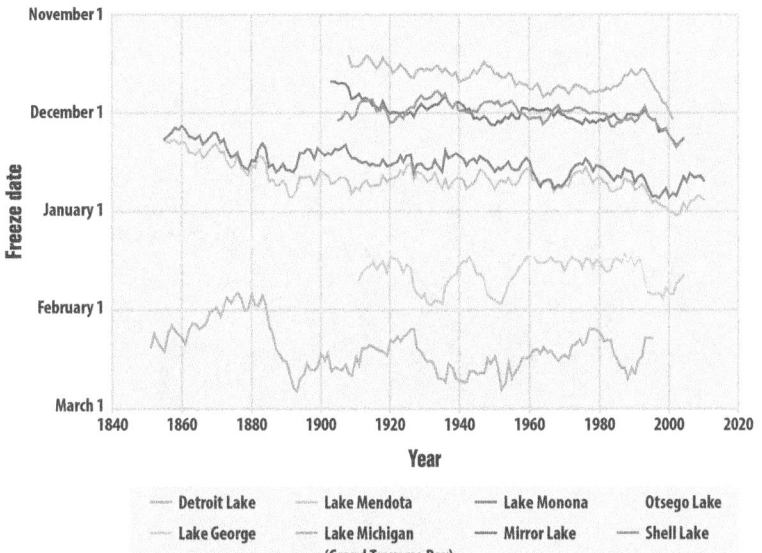

Data source: NSIDC, 2011[17]

Figure 3. Date of Ice Thaw for Selected U.S. Lakes, 1850–2010

This figure shows the "ice-off" date, or date of ice thawing and breakup, for eight U.S. lakes. The data are available from approximately 1850 to 2010, depending on the lake, and have been smoothed using a nine-year moving average.

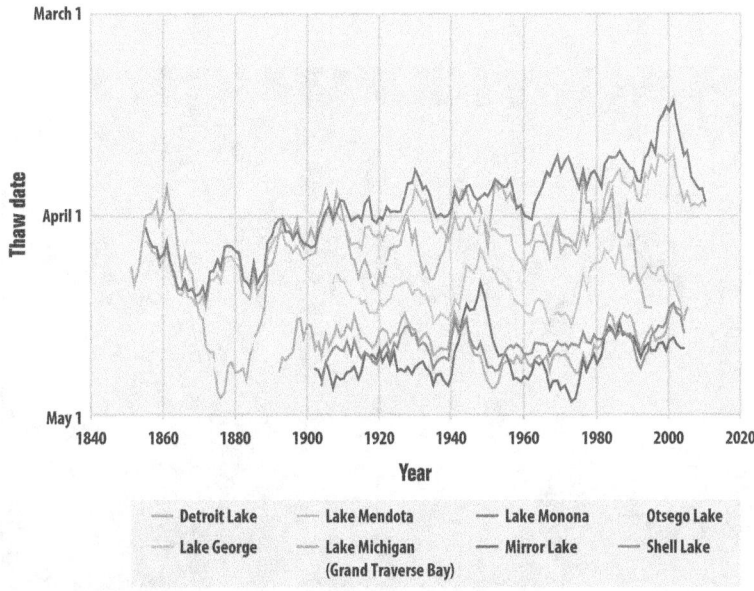

Data source: NSIDC, 2011[18]

cover breaks up and open water becomes extensive.

Freeze and thaw dates have been recorded through human visual observations for more than 150 years. The National Snow and Ice Data Center maintains a database with freeze and thaw observations from more than 700 lakes and rivers throughout the Northern Hemisphere. This indicator focuses on eight lakes within the United States that have the longest and most complete historical records. The lakes of interest are located in Minnesota, Wisconsin, Michigan, and New York.

Indicator Notes

Although there is a lengthy historical record of freeze and thaw dates for a much larger set of lakes and rivers, some records are incomplete, with breaks ranging from brief lapses to large gaps in data. This indicator is limited to eight lakes with fairly complete historical records.

Data used in this indicator are all based on visual observations. Records based on visual observations by individuals are open to some interpretation and can differ from one individual to the next. In addition, historical observations for lakes have typically been made from the shore, which might not be representative of lakes as a whole or comparable to more recent satellite-based observations.

Data Sources

Data were obtained from the Global Lake and River Ice Phenology Database, which is maintained by the National Snow and Ice Data Center. These data are available at: http://nsidc.org/data/lake_river_ice.

Snowfall

Background

Snowfall is an important aspect of winter in many parts of the United States. People depend on snow to provide water when it melts in the spring, and many communities rely on snow for winter recreation. Some plants and animals also depend on snow and snowmelt for survival. The amount of snow that falls in a particular area directly influences both snow cover and snowpack, which refer to snow that accumulates on the ground (see the Snow Cover indicator on p. 56 and the Snowpack indicator on p. 58).

Warmer temperatures cause more water to evaporate from the land and oceans, which leads to larger storms and more precipitation. In general, a warmer climate will cause more of this precipitation to fall in the form of rain instead of snow. However, some places could see more snowfall if temperatures rise but still remain below the freezing point, or if storm tracks change. Areas near large lakes might also experience more snowfall as lakes remain unfrozen for longer periods, allowing more water to evaporate. In contrast, other areas might experience less snowfall as a result of wintertime droughts.

Changes in the amount and timing of snowfall could affect the spawning of fish in the spring and the amount of water available for people to use in the spring and summer. Changes in snowfall could also affect winter recreation activities, like skiing, and the people who depend on these activities to make a living.

About the Indicator

This indicator tracks total snowfall as well as the percentage of precipitation that falls in the form of snow versus rain. These data were collected from hundreds of weather stations across the contiguous 48 states.

Total snowfall is determined by the height of snow that accumulates each day. These measured values commonly appear in weather reports (for example, a storm that deposits 10 inches of snow). Figure 1 shows how snowfall accumulation totals changed between

(Continued on page 55)

Figure 1. Change in Total Snowfall in the Contiguous 48 States, 1930–2007

This figure shows the average rate of change in total snowfall from 1930 to 2007 at 419 weather stations in the contiguous 48 states. Blue circles represent increased snowfall; red circles represent a decrease.

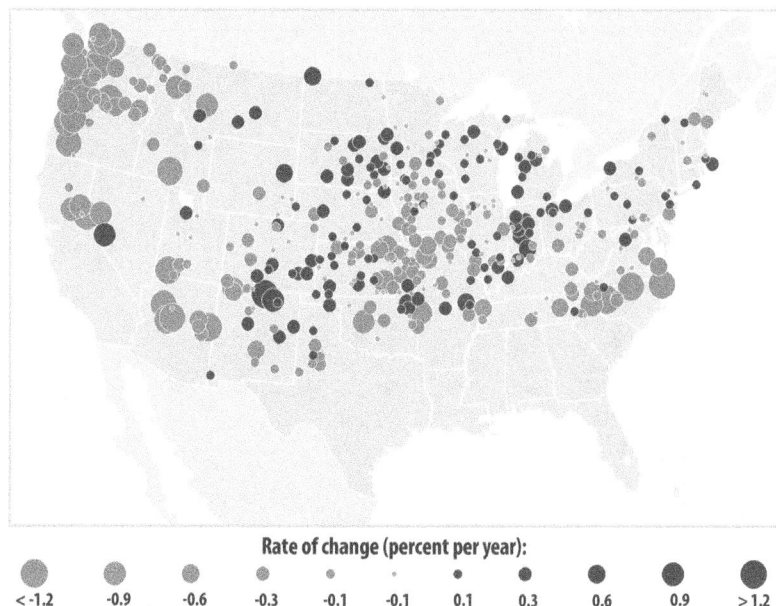

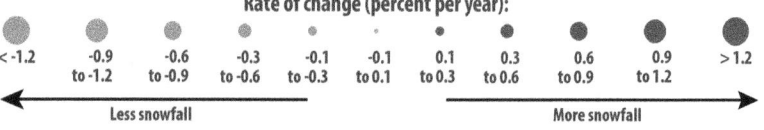

Data source: Kunkel et al., 2009[19]

Key Points

- Total snowfall has decreased in most parts of the country since widespread observations became available in 1930, with 57 percent of stations showing a decline (see Figure 1).

- In addition to changing the overall rate of precipitation, climate change can also lead to changes in the type of precipitation. One reason for the decline in total snowfall is because more winter precipitation is falling in the form of rain instead of snow. More than three-fourths of the stations across the contiguous 48 states have experienced a decrease in the proportion of precipitation falling as snow (see Figure 2).

- Snowfall trends vary by region. The Pacific Northwest has seen a decline in both total snowfall and the proportion of precipitation falling as snow. Parts of the Midwest have also experienced a decrease, particularly in terms of the snow-to-precipitation ratio. A few regions have seen modest increases, including some areas near the Great Lakes that now receive more snow than they used to (see Figures 1 and 2).

Figure 2. Change in Snow-to-Precipitation Ratio in the Contiguous 48 States, 1949–2011

This figure shows the percentage change in winter snow-to-precipitation ratio from 1949 to 2011 at 289 weather stations in the contiguous 48 states. This ratio measures what percentage of total winter precipitation falls in the form of snow. A decrease (red circle) indicates that more precipitation is falling in the form of rain instead of snow. Filled circles represent stations where the trend was statistically significant.

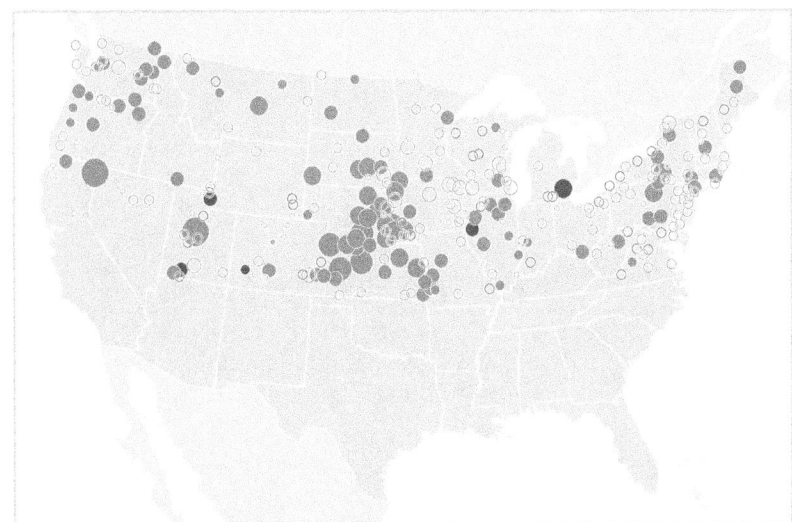

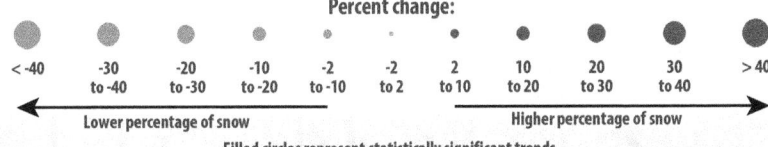

Filled circles represent statistically significant trends.
Open circles represent trends that are not statistically significant.

Data source: Feng, 2012[20]

1930 and 2007 at more than 400 weather stations. These stations were selected because they had high-quality data for the full timespan of this indicator.

Figure 2 of this indicator shows trends in the proportion of total precipitation that falls in the form of snow during each winter season. This is called the "snow-to-precipitation" ratio, and it is based on comparing the amount of snowfall with the total amount of precipitation (snow plus rain). For this comparison, snow has been converted to the equivalent amount of liquid water. These data are available from 1949 to 2011.

Indicator Notes

Several factors make it difficult to measure snowfall precisely. The snow accumulations shown in Figure 1 are based on the use of measuring rods. This measurement method is subject to human error, as well as the effects of wind (drifting snow) and the surrounding environment (such as tall trees). Similarly, snow gauges for Figure 2 may catch slightly less snow than rain because of the effects of wind. However, steps have been taken to limit this indicator to weather stations with the most consistent methods and the highest-quality data.[21] As a result, some parts of the country have a higher station density than others.

Both figures are limited to the winter season. Figure 1 comes from an analysis of October-to-May snowfall, while Figure 2 covers November through March. Although these months account for the vast majority of snowfall in most locations, this indicator might not represent the entire snow season in some areas.

Data Sources

This indicator shows trends based on two sets of weather records collected and maintained by the National Oceanic and Atmospheric Administration. Figure 1 was adapted from an analysis by Kunkel et al. (2009)[22] based on records from Cooperative Observer Program weather stations. Figure 2 is an updated version of an analysis by Feng and Hu (2007)[23] using data from the U.S. Historical Climatology Network. Additional information about the Cooperative Observer Program is available online at: www.nws.noaa.gov/om/coop. Information about the U.S. Historical Climatology Network can be found at: www.ncdc.noaa.gov/oa/climate/research/ushcn.

Snow Cover

Background

Snow cover refers to the amount of land covered by snow at any given time. Naturally, it is influenced by the amount of precipitation that falls as snow. Air temperature also plays a role because it determines whether precipitation falls as snow or rain, and it affects the rate at which snow on the ground will melt. As temperature and precipitation patterns change, so can the overall area covered by snow.

Snow cover is not just something that is affected by climate change; it also exerts an influence on climate. Because snow is white, it reflects much of the sunlight that hits it. In contrast, darker surfaces such as open water absorb more light and heat up more quickly. In this way, the overall amount of snow cover affects patterns of heating and cooling over the Earth's surface. More snow means more energy reflects back to space, while less snow cover means the Earth will absorb more heat and become warmer.

On a more local scale, snow cover is important for many plants and animals. For example, some plants rely on a protective blanket of snow to insulate them from sub-freezing winter temperatures. Humans and ecosystems also rely on snowmelt to replenish streams and ground water.

About the Indicator

This indicator tracks the total area covered by snow across all of North America (not including Greenland) since 1972. It is based on maps generated by analyzing satellite images collected by the National Oceanic and Atmospheric Administration. The indicator was created by analyzing each weekly map to determine the extent of snow cover, then averaging the weekly observations together to get a value for each year. Average snow cover was also calculated for each season: spring (defined as March–May), summer (June–August), fall (September–November), and winter (December–February). All maps were recently reanalyzed using the most precise methods available, making this the best available data set for assessing snow cover on a continental scale.

Figure 1. Snow-Covered Area in North America, 1972–2011

This graph shows the average area covered by snow in a given calendar year, based on an analysis of weekly maps. The area is measured in square miles. These data cover all of North America (not including Greenland).

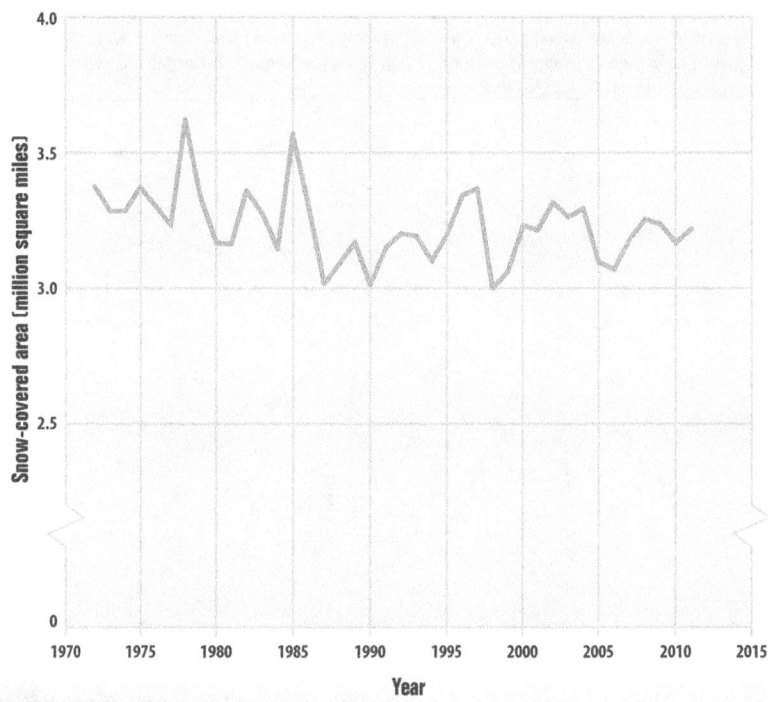

Data source: Rutgers University Global Snow Lab, 2012[24]

Figure 2. Snow-Covered Area in North America by Season, 1972–2011

This graph shows the average area covered by snow during spring (March–May), summer (June–August), fall (September–November), and winter (December–February), based on an analysis of weekly maps. The area is measured in square miles. These data cover all of North America (not including Greenland).

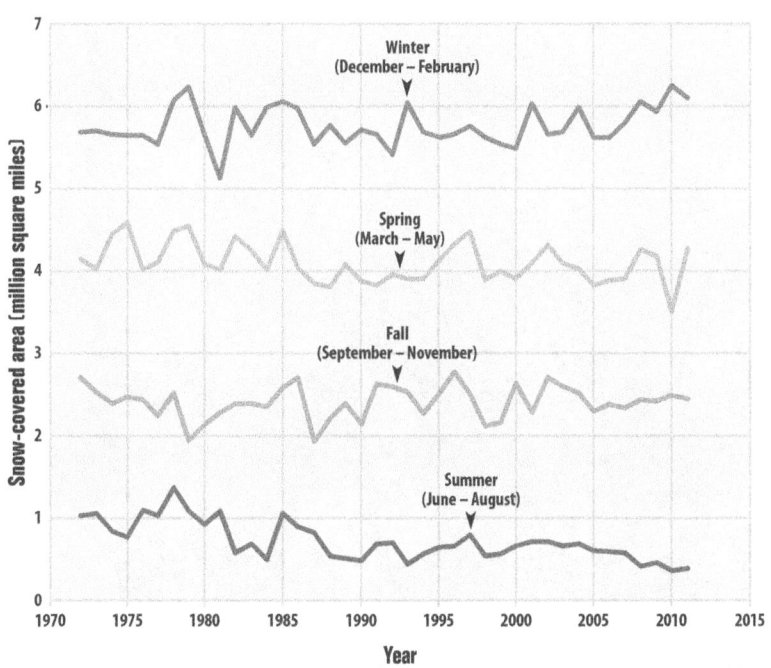

Data source: Rutgers University Global Snow Lab, 2012[25]

Indicator Notes

Although satellite-based snow cover maps are available starting in the mid-1960s, some of the early years are missing data from several weeks during the summer, which would lead to an inaccurate annual average. Thus, the indicator is restricted to 1972 and later, with all years having a full set of data.

Data Sources

The data for this indicator were provided by the Rutgers University Global Snow Lab, which posts data online at: http://climate.rutgers.edu/snowcover. The data are based on measurements collected by the National Oceanic and Atmospheric Administration's National Environmental Satellite, Data, and Information Service at: www.nesdis.noaa.gov.

Snowpack

This indicator measures trends in mountain snowpack in western North America.

Background

Temperature and precipitation are key factors affecting snowpack, which is the amount or thickness of snow that accumulates on the ground. In a warming climate, more precipitation will be expected to fall as rain rather than snow in most areas—reducing the extent and depth of snowpack. Higher temperatures in the spring can cause snow to melt earlier.

Mountain snowpack plays a key role in the water cycle in western North America, storing water in the winter when the snow falls and releasing it as runoff in spring and summer when the snow melts. Millions of people in the West depend on the melting of mountain snowpack for power, irrigation, and drinking water. In most western river basins, snowpack is a larger component of water storage than human-constructed reservoirs.[26]

Changes in mountain snowpack can affect agriculture, winter recreation, and tourism in some areas, as well as plants and wildlife. For example, certain types of trees rely on snow for insulation from freezing temperatures, as do some animal species. In addition, fish spawning could be disrupted if changes in snowpack or snowmelt alter the timing and abundance of streamflows.

About the Indicator

This indicator uses a measurement called snow water equivalent to determine trends in snowpack. Snow water equivalent is the amount of water contained within the snowpack at a particular location. It can be thought of as the depth of water that would result if the entire snowpack were to melt.

The U.S. Department of Agriculture and other collaborators have measured snowpack since the 1930s. In the early years of data collection, researchers measured snow water equivalent manually, but since 1980, measurements at some locations have been collected with automated instruments. This indicator is based on data from approximately 800 permanent research sites in the western United States

(Continued on page 59)

Figure 1. Trends in April Snowpack in the Western United States and Canada, 1950–2000

This map shows trends in April snowpack in the western United States and part of Canada, measured in terms of snow water equivalent. Blue circles represent increased snowpack; red circles represent a decrease.

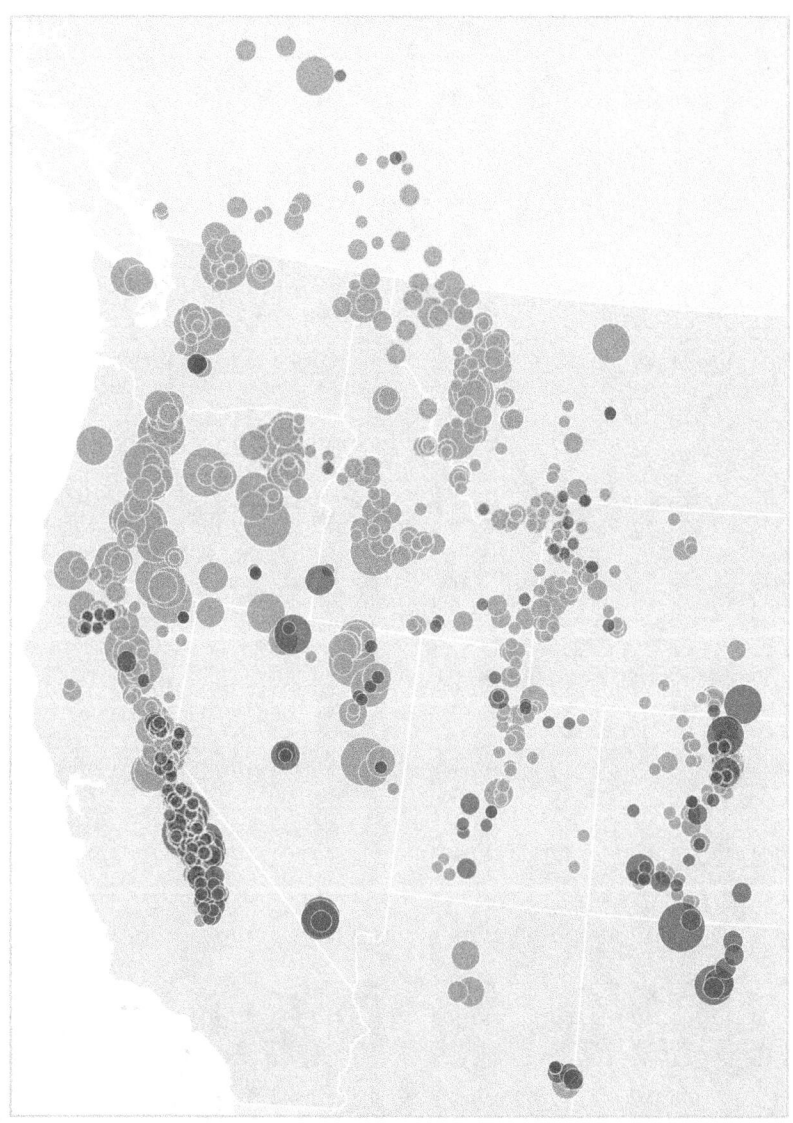

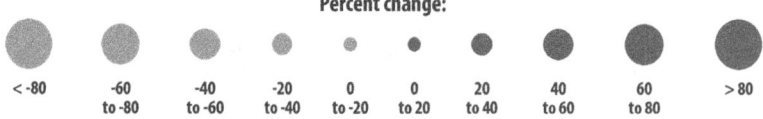

Percent change:

| < -80 | -60 to -80 | -40 to -60 | -20 to -40 | 0 to -20 | 0 to 20 | 20 to 40 | 40 to 60 | 60 to 80 | > 80 |

Data source: Mote, 2009[27]

Key Points

- From 1950 to 2000, April snowpack declined at most of the measurement sites (see Figure 1), with some relative losses exceeding 75 percent.

- In general, the largest decreases were observed in western Washington, western Oregon, and northern California. April snowpack decreased to a lesser extent in the northern Rockies.

- A few areas have seen increases in snowpack, primarily in the southern Sierra Nevada of California and in the Southwest.

and Canada. The indicator shows long-term rates of change for the month of April, which could reflect changes in winter snowfall as well as the timing of spring snowmelt.

Indicator Notes

Natural changes in the Earth's climate could affect snowpack in such a way that trends might slightly differ if measured over a different time period. The 1950s registered some of the highest snowpack measurements of the 20th century in the Northwest. While these values could be magnifying the extent of the snowpack decline depicted in Figure 1, the general direction of the trend is the same regardless of the start date.

Although most parts of the West have seen reductions in snowpack consistent with overall warming trends shown in the U.S. and Global Temperature indicator (p. 24) snowfall trends may be partially influenced by nonclimatic factors such as observation methods, land-use changes, and forest canopy changes.

Data Sources

Data for this indicator came from the U.S. Department of Agriculture's Natural Resources Conservation Service Water and Climate Center. The map was constructed using methods described in Mote et al. (2005).[28] The U.S. Department of Agriculture data are available at: www.wcc.nrcs.usda.gov.

Society and

INDICATORS IN THIS CHAPTER

 **Streamflow**

 **Ragweed Pollen Season**

 **Length of Growing Season**

Ecosystems

The indicators in this report show that changes are occurring throughout the Earth's climate system, including increases in air and ocean temperatures, more extreme weather events, a rise in sea level, widespread melting of glaciers, and longer ice-free periods on lakes and rivers. Changes such as these can present a wide range of challenges to human well-being, the economy, and natural ecosystems.

What is happening?

For society, human health effects from increases in temperature are likely to include increases in heat-related illnesses and deaths, especially in urban areas. Changes in precipitation patterns and timing affect streamflow and water availability, while more severe storms and floods damage property and infrastructure (such as roads, bridges, and utilities) and cause loss of life. Warming temperatures are also affecting the length of the growing season in the United States as the timing between the last (spring) and first (fall) frost has expanded by nearly two weeks over the last 100 years.

While species have adapted to environmental change for millions of years, climate change could require adaptation on larger and faster scales than current species have successfully achieved in the past.

Changes in climate can affect ecosystems by influencing animal behavior, such as nesting and migration patterns, as well as the timing and extent of natural processes such as flower blooms and the length of pollen seasons in many areas.

Why does it matter?

Ecosystems provide humans with food, clean water, and a variety of other services that can be affected by climate change. While species have adapted to environmental change for millions of years, the climate changes being experienced now could require adaptation on larger and faster scales than current species have successfully achieved in the past, thus increasing the risk of extinction for some species.

The more the climate changes, the greater the potential effects on society and ecosystems. The nature and extent of climate change effects, and whether these effects will be harmful or beneficial, will vary regionally and over time. The extent to which climate change will affect different ecosystems, regions, and sectors of society will depend not only on the sensitivity of those systems to climate change, but also on their ability to adapt to or cope with climate change.

For detailed information about data used in these indicators, see the online technical documentation at: www.epa.gov/climatechange/indicators.

Leaf and Bloom Dates

Bird Wintering Ranges

Heat-Related Deaths

Streamflow

Background

Streamflow is a measure of the amount of water carried by rivers and streams, and it represents a critical resource for people and the environment. Changes in streamflow can directly influence the supply of drinking water and the amount of water available for irrigating crops, generating electricity, and other needs. In addition, many plants and animals depend on streamflow for habitat and survival.

Streamflow naturally varies over the course of a year. For example, rivers and streams in many parts of the country have their highest sustained flow when snow melts in the spring. The amount of streamflow is important because very high flows can cause erosion and damaging floods, while very low flows can diminish water quality, harm fish, and reduce the amount of water available for people to use. The timing of peak flow is important because it affects the ability of reservoir managers to store water to meet people's needs later in the year. In addition, some plants and animals (such as fish that migrate) depend on a particular pattern of streamflow as part of their life cycles.

Climate change can affect streamflow in several ways. Changes in the amount of snowpack and earlier spring melting (see the Snowpack indicator on p. 58) can alter the size and timing of peak streamflows. More precipitation is expected to cause higher average streamflow in some places, while heavier storms (see the Heavy Precipitation indicator on p. 30) could lead to larger peak flows. More frequent or severe droughts will reduce streamflow in certain areas.

About the Indicator

The U.S. Geological Survey measures streamflow in rivers and streams across the United States using continuous monitoring devices called stream gauges. This indicator is based on 211 stream gauges located in areas where trends will not be artificially influenced by dams, reservoir management, wastewater treatment facilities, or other activities.

(Continued on page 63)

Figure 1. Volume of Seven-Day Low Streamflows in the United States, 1940–2009

This map shows percentage changes in the minimum amount of water carried by rivers and streams across the country, based on the long-term rate of change from 1940 to 2009. Minimum streamflow is based on the seven-day period with the lowest average flow during a given year.

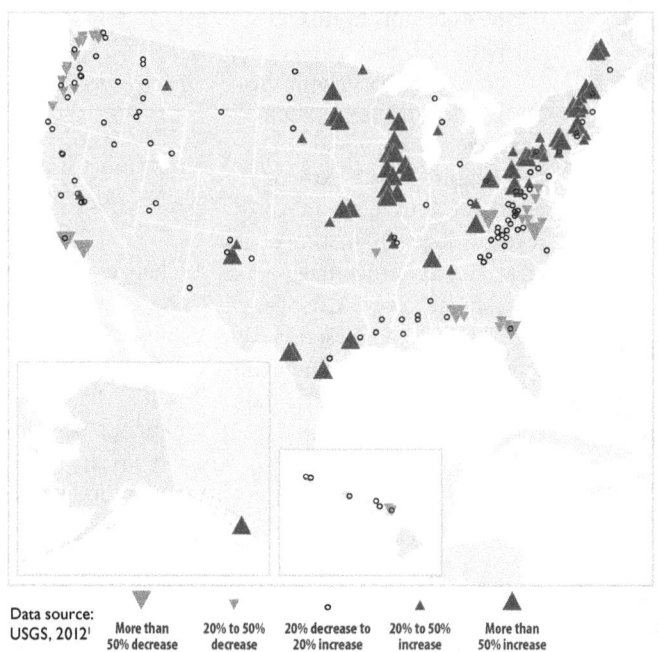

Data source: USGS, 2012[1]

| More than 50% decrease | 20% to 50% decrease | 20% decrease to 20% increase | 20% to 50% increase | More than 50% increase |

Figure 2. Volume of Three-Day High Streamflows in the United States, 1940–2009

This map shows percentage changes in the maximum amount of water carried by rivers and streams across the country, based on the long-term rate of change from 1940 to 2009. Maximum streamflow is based on the three-day period with the highest average flow during a given year.

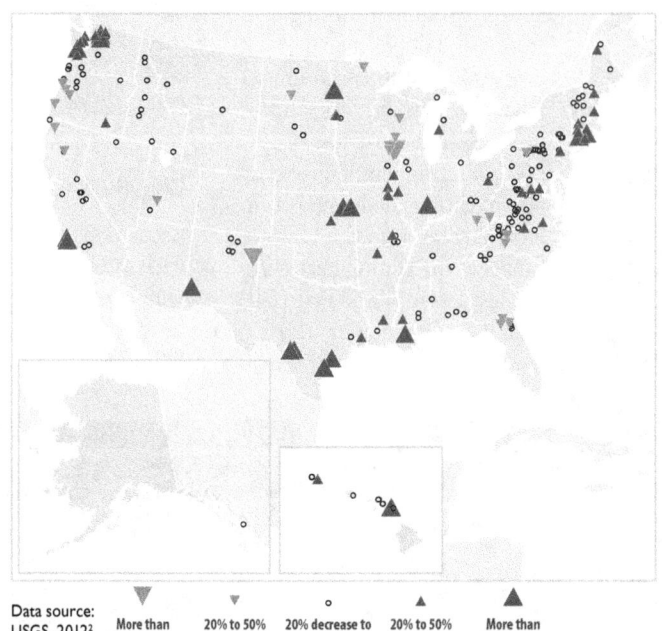

Data source: USGS, 2012[2]

| More than 50% decrease | 20% to 50% decrease | 20% decrease to 20% increase | 20% to 50% increase | More than 50% increase |

Key Points

- Over the past 70 years, seven-day low flows have generally increased in the Northeast and Midwest (in other words, on the driest days, streams are carrying more water than before). Low flows have generally decreased (that is, streams are carrying less water than before) in parts of the Southeast and the Pacific Northwest. Overall, more sites have seen increases than decreases (see Figure 1).

- Three-day high-flow trends vary from region to region across the country. For example, streams in the Northeast have generally seen an increase or little change in high flows since 1940, while some West Coast streams have seen a decrease and others have seen an increase (see Figure 2). Overall, more sites have seen increases than decreases.

- Sixty percent of the streams measured show winter-spring runoff happening more than five days earlier than it did in the past. The most dramatic change has occurred in the Northeast (see Figure 3).

Figure 3. Timing of Winter-Spring Runoff in the United States, 1940–2009

This map shows changes in the timing of peak spring flow carried by rivers and streams, based on the long-term rate of change from 1940 to 2009. This analysis focuses on parts of the country where streamflow is strongly influenced by snowmelt. It is based on the winter-spring center of volume, which is the date when half of the streamflow between January 1 and May 31 of each year has passed.

▽	▽	▽	○	▲	▲	▲
More than 10 days earlier	5 to 10 days earlier	2 to 5 days earlier	2 days earlier to 2 days later	2 to 5 days later	5 to 10 days later	More than 10 days later

Data source: USGS, 2012[3]

This indicator examines three important measures of streamflow conditions that occur over the course of a year. Figure 1 looks at the driest conditions each year, which are commonly calculated by averaging the lowest seven consecutive days of streamflow over the year. This method captures the year's most severe, sustained dry spell. Figure 2 examines high flow conditions, which are commonly calculated as the highest average flow over three consecutive days. Based on typical weather patterns, three days is an optimal length of time to capture runoff associated with large storms and peak snowmelt.

Figure 3 shows changes in the timing of spring runoff over time. This measure is limited to 55 stream gauges in areas where at least 30 percent of annual precipitation falls as snow. Scientists look at the total volume of water that passes by a gauge between January 1 and May 31 and then determine the date when exactly half of the water has gone by. This date is called the winter-spring center of volume. A long-term trend toward an earlier date suggests that spring snowmelt is happening earlier.

Indicator Notes

Measurements were taken in areas where streamflow is not highly affected by human influences, including changes in land cover. However, changes in land cover and land use over time could still influence streamflow trends at some streams. The gauges used for this indicator are not evenly distributed across the country.

Data Sources

Streamflow data were collected by the U.S. Geological Survey. These data came from a set of watersheds with minimal human impacts, which have been classified as reference gauges.[4] Daily average streamflow data are stored in the National Water Information System and are publicly available at: http://waterdata.usgs.gov/nwis.

Ragweed Pollen Season

This indicator depicts changes in the length of ragweed pollen season in the United States and Canada.

Background

More than half of Americans have at least one allergy.[5] Allergies are a major public health concern, with hay fever (congestion, runny nose, itchy eyes) accounting for more than 13 million visits to physicians' offices and other medical facilities every year.[6] One of the most common environmental allergens is ragweed, which can cause hay fever and trigger asthma attacks, especially in children and the elderly. An estimated 26 percent of all Americans are sensitive to ragweed.[7]

Ragweed plants mature in mid-summer and produce small flowers that generate pollen. Ragweed pollen season usually peaks in late summer and early fall, but these plants often continue to produce pollen until the first frost. A single ragweed plant can produce up to a billion pollen grains in one season, and these grains can be carried long distances by the wind.[8]

Climate change can affect pollen allergies in several ways. Warmer spring temperatures cause some plants to start producing pollen earlier (see the Leaf and Bloom Dates indicator on p. 68), while warmer fall temperatures extend the growing season for other plants such as ragweed (see the Length of Growing Season indicator on p. 66). Warmer temperatures and increased carbon dioxide concentrations also enable ragweed and other plants to produce more pollen.[9] This means that many locations could experience longer allergy seasons and higher pollen counts as a result of climate change.

About the Indicator

This indicator shows changes in the length of the ragweed pollen season in 10 cities in the central United States and Canada. These locations were selected as part of a study that looked at trends in pollen season at sites similar in elevation but across a range of latitudes from south to north. At each location, air

(Continued on page 65)

Figure 1. Change in Ragweed Pollen Season, 1995–2011

This figure shows how the length of ragweed pollen season changed at 10 locations in the central United States and Canada between 1995 and 2011. Red circles represent a longer pollen season; blue circles represent a shorter season. Larger circles indicate larger changes.

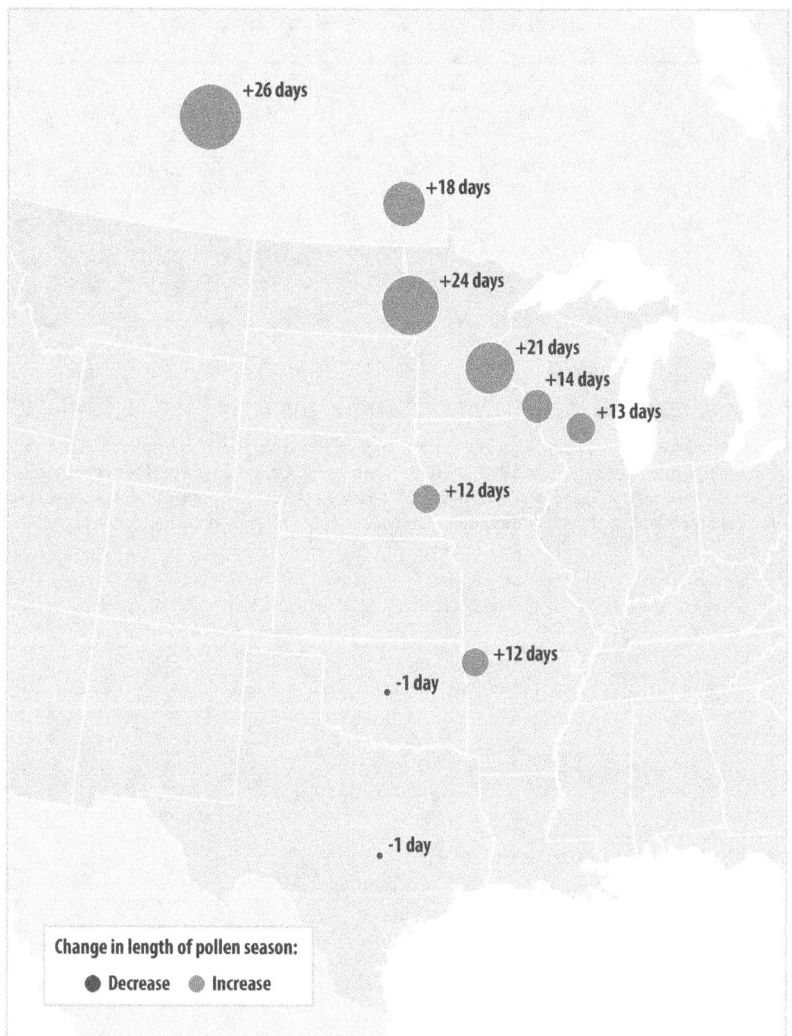

Data source: Ziska et al., 2012[10]

Key Points

- Since 1995, ragweed pollen season has grown longer at eight of the 10 locations studied (see Figure 1).

- The increase in ragweed season length becomes more pronounced from south to north. Ragweed season increased by 24 days in Fargo, North Dakota, and 26 days in Saskatoon, Saskatchewan (see Figure 1). This trend is consistent with many other observations showing that climate is changing more rapidly at higher latitudes.[11]

- The trends in Figure 1 are strongly related to changes in the length of the frost-free season and the timing of the first fall frost. Northern areas have seen fall frosts happening later than they used to, with the delay in first frost closely matching the increase in pollen season. Meanwhile, some southern stations have experienced only a modest change in frost-free season length since 1995.[12]

samples have been collected and examined for at least 16 years as part of a national allergy monitoring network. Pollen spores are counted and identified using microscopes.

Pollen counts from each station have been analyzed to determine the start and end dates of each year's ragweed pollen season. Because the length of ragweed season naturally varies from year to year, statistical techniques have been used to determine the average rate of change over time. This indicator shows the total change in season length from 1995 to 2011, which was determined by multiplying the average annual rate of change by the number of years in the period.

Indicator Notes

This indicator is based on data from a limited number of cities in the central states and provinces. These cities cover a broad range from north to south, however, which allows researchers to establish a clear connection between pollen season changes and latitude.

Many factors can influence year-to-year changes in pollen season, including typical variations in temperature and precipitation, extreme events such as floods and droughts, and changes in plant diversity. Adding more years of data would provide a better picture of long-term trends, but widespread data were not available prior to 1995.

This indicator does not show how the intensity of ragweed pollen season (pollen counts) might also be changing.

Data Sources

Data for this indicator come from the National Allergy Bureau, which is part of the American Academy of Allergy, Asthma, and Immunology's Aeroallergen Network. Data were compiled and analyzed by a team of researchers who published a more detailed version of this analysis in a scientific journal with data through 2009.[13]

Length of Growing Season

This indicator measures the length of the growing season in the contiguous 48 states.

Background

The length of the growing season in any given region refers to the number of days when plant growth takes place. The growing season often determines which crops can be grown in an area, as some crops require long growing seasons, while others mature rapidly. Growing season length is limited by many different factors. Depending on the region and the climate, the growing season is influenced by air temperatures, frost days, rainfall, or daylight hours.

Changes in the length of the growing season can have both positive and negative effects. Moderate warming can benefit crop and pasture yields in mid- to high-latitude regions, yet even slight warming decreases yields in seasonally dry and low-latitude regions.[14] A longer growing season could allow farmers to diversify crops or have multiple harvests from the same plot. However, it could also limit the types of crops grown, encourage invasive species or weed growth, or increase demand for irrigation. A longer growing season could also disrupt the function and structure of a region's ecosystems and could, for example, alter the range and types of animal species in the area.

About the Indicator

This indicator looks at the impact of temperature on the length of the growing season in the contiguous 48 states, as well as trends in the timing of spring and fall frosts. For this indicator, the length of the growing season is defined as the period of time between the last frost of spring and the first frost of fall, when the air temperature drops below the freezing point of 32°F.

Trends in the growing season were calculated using temperature data from 750 weather stations throughout the contiguous 48 states. These data were obtained from the National Oceanic and Atmospheric Administration's National Climatic Data Center. Growing season length and the timing of spring and fall frosts were averaged across the nation, then compared with long-term average numbers (1895–2011) to determine how each year differed from the long-term average.

Figure 1. Length of Growing Season in the Contiguous 48 States, 1895–2011

This figure shows the length of the growing season in the contiguous 48 states compared with a long-term average. For each year, the line represents the number of days shorter or longer than average. The line was smoothed using an 11-year moving average. Choosing a different long-term average for comparison would not change the shape of the data over time.

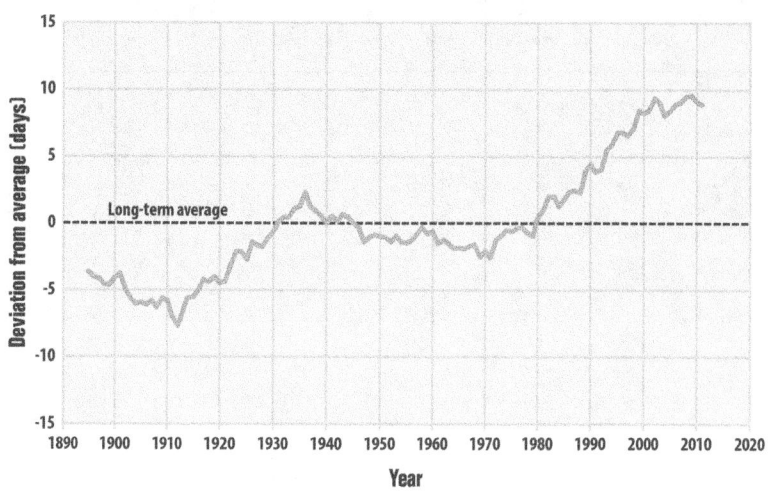

Data source: Kunkel, 2012[15]

Figure 2. Length of Growing Season in the Contiguous 48 States, 1895–2011: West Versus East

This figure shows the length of the growing season in the western and eastern United States compared with a long-term average. For each year, the line represents the number of days shorter or longer than average. The lines were smoothed using an 11-year moving average. Choosing a different long-term average for comparison would not change the shape of the data over time.

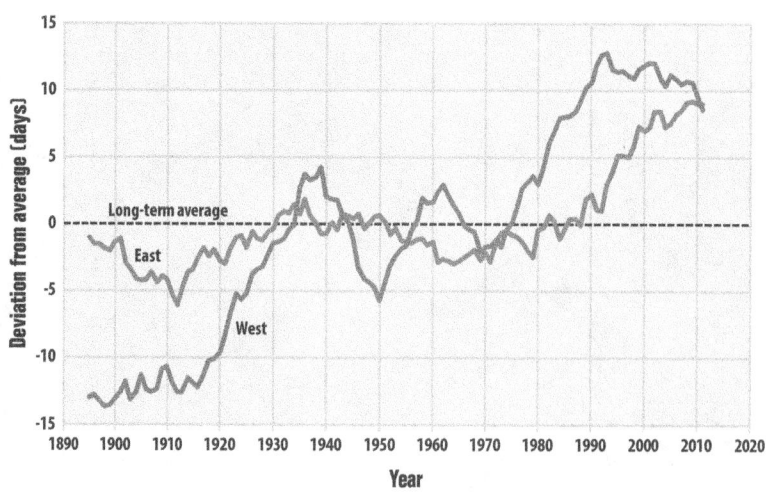

Data source: Kunkel, 2012[16]

Key Points

- The average length of the growing season in the contiguous 48 states has increased by nearly two weeks since the beginning of the 20th century. A particularly large and steady increase occurred over the last 30 years (see Figure 1).

- The length of the growing season has increased more rapidly in the West than in the East. In the West, the length of the growing season has increased at an average rate of about 22 days per century since 1895, compared with a rate of about eight days per century in the East (see Figure 2).

- The final spring frost is now occurring earlier than at any point since 1895, and the first fall frosts are arriving later. Since 1980, the last spring frost has occurred an average of about three days earlier than the long-term average, and the first fall frost has occurred about two days later (see Figure 3).

Indicator Notes

Changes in measurement techniques and instruments over time can affect trends. This indicator only includes data from weather stations with a consistent record of data points for the time period.

Data Sources

All three figures are based on temperature data compiled by the National Oceanic and Atmospheric Administration's National Climatic Data Center, and these data are available online at: www.ncdc.noaa.gov/oa/ncdc.html. Frost timing and growing season length were analyzed by Kunkel (2012).[18]

Figure 3. Timing of Last Spring Frost and First Fall Frost in the Contiguous 48 States, 1895–2011

This figure shows the timing of the last spring frost and the first fall frost in the contiguous 48 states compared with a long-term average. Positive values indicate that the frost occurred later in the year, and negative values indicate that the frost occurred earlier in the year. The lines were smoothed using an 11-year moving average. Choosing a different long-term average for comparison would not change the shape of the data over time.

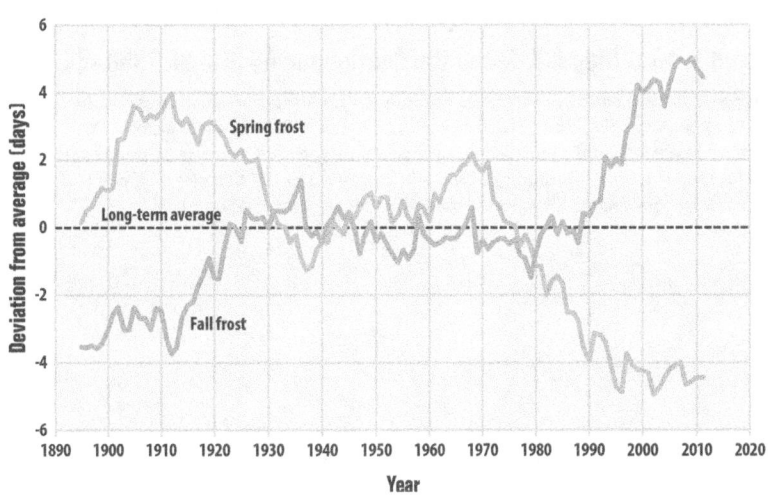

Data source: Kunkel, 2012[17]

Leaf and Bloom Dates

Background

The timing of natural events, such as flower blooms and animal migration, is influenced by changes in climate. Phenology is the study of such important seasonal events. Phenological events are influenced by a combination of climate factors, including light, temperature, rainfall, and humidity. Different plant and animal species respond to different cues.

Scientists have very high confidence that the earlier arrival of spring events is linked to recent warming trends in global climate.[19] Disruptions in the timing of these events can have a variety of impacts on ecosystems and human society. For example, an earlier spring might lead to longer growing seasons (see the Length of Growing Season indicator on p. 66), more abundant invasive species and pests, and earlier and longer allergy seasons.

Because of their close connection with climate, the timing of phenological events can be used as an indicator of the sensitivity of ecological processes to climate change. Some phenological indicators cover broad trends, such as overall "leaf-on" dates (when trees grow new leaves in the spring), using a combination of satellite data and ground observations. Others rely on ground observations that look at specific types or species of plants or animals. Two particularly useful indicators of the timing of spring events are the first leaf dates and the first bloom dates of lilacs and honeysuckles, which have an easily monitored flowering season, relatively high survival rate, and large geographic distribution. The first leaf date in these plants relates to the timing of "early spring," while the first bloom date is consistent with the timing of later spring events, such as the start of growth in forest vegetation.[20]

About the Indicator

This indicator shows trends in the timing of first leaf dates and first bloom dates in lilacs and honeysuckles across the contiguous 48 states. Because many of the phenological observation records in the United States are less than 40 years long, and because these records may have

(Continued on page 69)

Figure 1. First Leaf Dates in the Contiguous 48 States, 1900–2010

This figure shows modeled trends in lilac and honeysuckle first leaf dates across the contiguous 48 states, using the 1981 to 2010 average as a baseline. Positive values indicate that leaf growth began later in the year, and negative values indicate that leafing occurred earlier. The thicker line was smoothed using a nine-year weighted average. Choosing a different long-term average for comparison would not change the shape of the data over time.

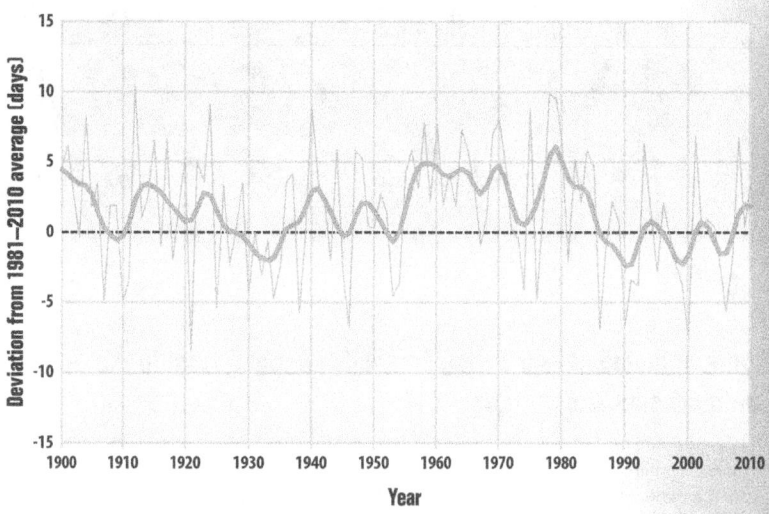

Data source: Schwartz, 2011[21]

Figure 2. First Bloom Dates in the Contiguous 48 States, 1900–2010

This figure shows modeled trends in lilac and honeysuckle first bloom dates across the contiguous 48 states, using the 1981 to 2010 average as a baseline. Positive values indicate that blooming began later in the year, and negative values indicate that blooming occurred earlier. The thicker line was smoothed using a nine-year weighted average. Choosing a different long-term average for comparison would not change the shape of the data over time.

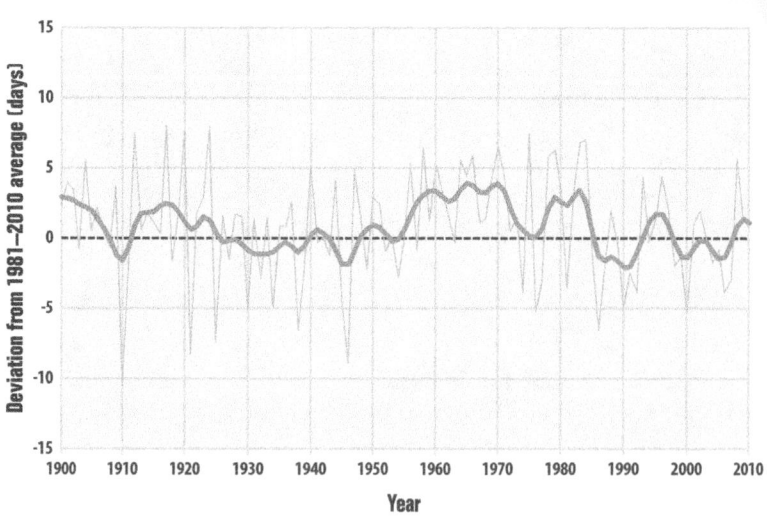

Data source: Schwartz, 2011[22]

gaps in time or space, computer models have been used to provide a more complete understanding of long-term trends nationwide.

The models for this indicator were developed using data from the USA National Phenology Network, which collects ground observations from a network of federal agencies, field stations, educational institutions, and citizens who have been trained to log observations of leaf and bloom dates. For consistency, observations were limited to a few specific types of lilacs and honeysuckles. Next, models were created to relate actual leaf and bloom observations with records from nearby weather stations. Once scientists were able to determine the relationship between leaf and bloom dates and climate factors (particularly temperatures), they used this knowledge to estimate leaf and bloom dates for earlier years based on historical weather records. They also used the models to estimate how leaf and bloom dates would have changed in a few areas (mostly in the far South) where lilacs and honeysuckles are not widespread.

This indicator uses data from several hundred weather stations throughout the contiguous 48 states. The exact number of stations varies from year to year. For each year, the timing of first leaf and first bloom at each station was compared with the 1981 to 2010 average to determine the number of days' "deviation from normal." This indicator presents the average deviation across all stations.

Indicator Notes

Plant phenological events are studied using several data collection methods, including satellite images, models, and direct observations. The use of varying data collection methods in addition to different phenological indicators (such as leaf or bloom dates for different types of plants) can lead to a range of estimates of the arrival of spring.

Climate is not the only factor that can affect phenology. Observed variations can also reflect plant genetics, changes in the surrounding ecosystem, and other factors. This indicator minimizes genetic influences by relying on cloned plant species (that is, plants with no genetic differences).

Data Sources

Leaf and bloom observations were compiled by the USA National Phenology Network and are available at: www.usanpn.org. This indicator is also based on climate data that were provided by the U.S. Historical Climatology Network and are available at: www.ncdc.noaa.gov/oa/climate/research/ushcn. Data for this indicator were analyzed using methods described by McCabe et al. (2011).[24]

Key Points

- First leaf growth in lilacs and honeysuckles in the contiguous 48 states is now occurring a few days earlier than it did in the early 1900s. Although the data show a great deal of year-to-year variability, a noticeable change toward earlier dates seems to have begun around the 1980s (see Figure 1).

- Lilac and honeysuckle bloom dates show a high degree of year-to-year variability, which makes it difficult to determine whether a statistically meaningful change has taken place (see Figure 2).

- Other studies have looked at trends in leaf and bloom dates across all of North America and the entire Northern Hemisphere. These studies have also found a trend toward earlier spring events—some more pronounced than the trends seen in just the contiguous 48 states.[23]

Bird Wintering Ranges

Background

Changes in climate can affect ecosystems by influencing animal behavior and distribution. Birds are a particularly good indicator of environmental change for several reasons:

- Each species of bird has adapted to certain habitat types, food sources, and temperature ranges. In addition, the timing of certain events in their life cycles—such as migration and reproduction—is driven by cues from the environment. For example, many North American birds follow a regular seasonal migration pattern, moving north to feed and breed in the summer, then moving south to spend the winter in warmer areas. Changing conditions can influence the distribution of both migratory and nonmigratory birds as well as the timing of important life cycle events.

- Birds are easy to identify and count, and thus there is a wealth of scientific knowledge about their distribution and abundance. People have kept detailed records of bird observations for more than a century.

- There are many different species of birds living in a variety of habitats, including water birds, coastal birds, and land birds. If a change in habitats or habits occurs across a range of bird types, it suggests that a common force might be contributing to that change.

Temperature and precipitation patterns are changing across the United States (see the U.S. and Global Temperature indicator on p. 24 and the U.S. and Global Precipitation indicator on p. 28). Some bird species can adapt to generally warmer temperatures by changing where they live—for example, by migrating further north in the summer but not as far south in the winter, or by shifting inland as winter temperature extremes grow less severe. Nonmigratory species might shift as well, expanding into newly suitable habitats while moving out of areas that become less suitable. Other types of birds might not adapt to changing conditions and could experience a population decline as a result. Climate change can also alter the timing of events that are based on temperature cues, such as migration and breeding (especially egg-laying).

Figure 1. Change in Latitude of Bird Center of Abundance, 1966–2005

This figure shows annual change in latitude of bird center of abundance for 305 widespread bird species in North America from 1966 to 2005. Each winter is represented by the year in which it began (for example, winter 2005–2006 is shown as 2005). The shaded band shows the likely range of values, based on the number of measurements collected and the precision of the methods used.

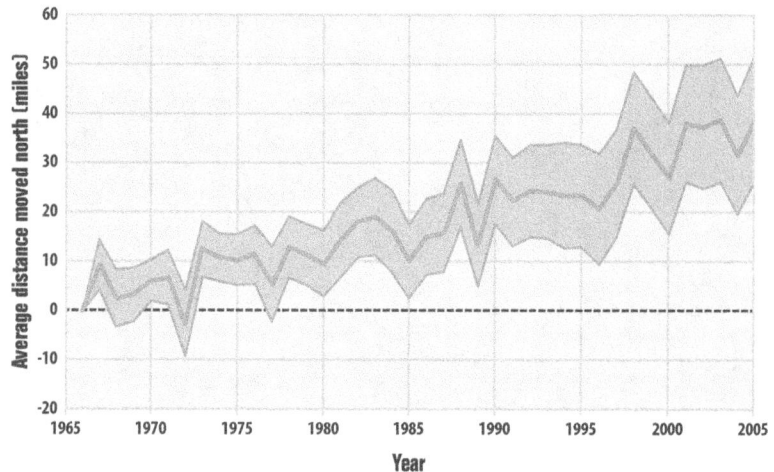

Data source: National Audubon Society, 2009[25]

Key Points

- Among 305 widespread North American bird species, the average mid-December to early January center of abundance moved northward between 1966 and 2005. The average species shifted northward by 35 miles during this period (see Figure 1). Trends in center of abundance are closely related to winter temperatures.[26]

- On average, bird species have also moved their wintering grounds farther from the coast since the 1960s (see Figure 2). This shift also relates to changes in winter temperatures.[27]

- Some species have moved farther than others. Of the 305 species studied, 177 (58 percent) have shifted their wintering grounds significantly northward since the 1960s, but some others have not moved at all. A few species have moved northward by as much as 200 to 400 miles.[28]

About the Indicator

This indicator looks at the "center of abundance" of 305 widespread North American bird species over a 40-year period. The center of abundance is a point on the map that represents the middle of each species' distribution. If a whole population of birds were to shift generally northward, one would see the center of abundance shift northward as well.

For year-to-year consistency, this indicator uses observations from the National Audubon Society's Christmas Bird Count, which takes place every year in early winter. The Christmas Bird Count is a long-running citizen science program in which individuals are organized by the National Audubon Society, Bird Studies Canada, local Audubon chapters, and other bird clubs to identify and count bird species. The data presented in this indicator were collected from more than 2,000 locations throughout the United States and parts of Canada. At each location, skilled observers follow a standard counting procedure to estimate the number of birds within a 15-mile diameter "count circle" over a 24-hour period. Study methods remain generally consistent from year to year. Data produced by the Christmas Bird Count go through several levels of review before Audubon scientists analyze the final data, which have been used to support a wide variety of peer-reviewed studies.

Figure 2. Change in Distance to Coast of Bird Center of Abundance, 1966–2005

This figure shows annual change in distance to the coast of bird center of abundance for 305 widespread bird species in North America from 1966 to 2005. Each winter is represented by the year in which it began (for example, winter 2005–2006 is shown as 2005). The shaded band shows the likely range of values, based on the number of measurements collected and the precision of the methods used.

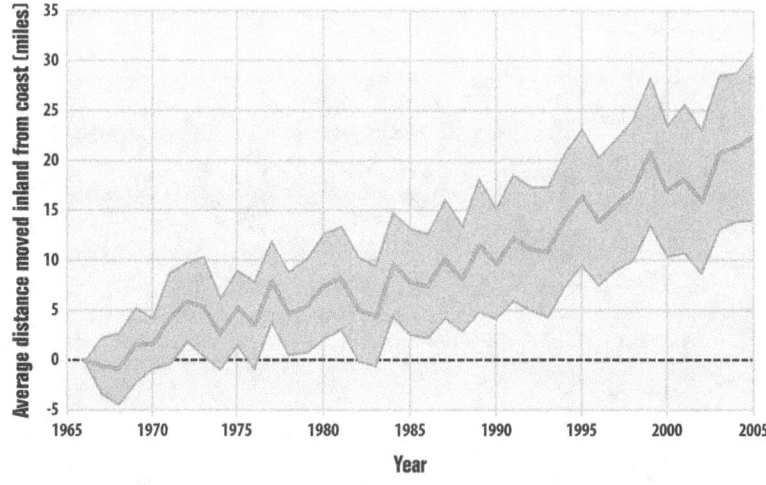

Data source: National Audubon Society, 2009[29]

Indicator Notes

Many factors can influence bird ranges, including food availability, habitat alteration, and interactions with other species. As a result, some of the birds included in this indicator might have moved north for reasons other than changing temperatures. This indicator does not show how responses to climate change vary among different types of birds. For example, a more detailed National Audubon Society analysis found large differences among coastal birds, grassland birds, and birds adapted to feeders, which all have varying abilities to adapt to temperature changes.[30]

Some data variations can be caused by differences among count circles, such as inconsistent level of effort by volunteer observers, but these differences are carefully corrected in Audubon's statistical analysis.

Data Sources

Bird center of abundance data were collected by the annual Christmas Bird Count organized by the National Audubon Society and Bird Studies Canada. Recent and historical Christmas Bird Count data are available at: http://birds.audubon.org/christmas-bird-count. Data for this indicator were analyzed by the National Audubon Society in 2009[31] and are available at: www.audubon.org/bird/bacc/index.html.

Heat-Related Deaths

Background

When people are exposed to extreme heat, they can suffer from potentially deadly heat-related illnesses such as heat exhaustion and heat stroke. Heat is the leading weather-related killer in the United States, even though most heat-related deaths are preventable through outreach and intervention (see EPA's Excessive Heat Events Guidebook at: www.epa.gov/heatisland/about/pdf/EHEguide_final.pdf)

Unusually hot summer temperatures have become more frequent across the contiguous 48 states in recent decades[32] (see the High and Low Temperatures indicator on p. 26), and extreme heat events (heat waves) are expected to become longer, more frequent, and more intense in the future.[33] As a result, the risk of heat-related deaths and illness is also expected to increase.[34]

Increases in summertime temperature variability may increase the risk of heat-related death for the elderly and other vulnerable populations.[35] Older adults carry the highest risk of heat-related death, although young children are also sensitive to the effects of heat. Across North America, the population over the age of 65 is growing dramatically as the baby boomer generation ages. People with certain diseases, such as cardiovascular and respiratory illnesses, are especially vulnerable to excessive heat exposure, as are the economically disadvantaged.

Some studies suggest that the number of deaths caused by extremely cold temperatures might drop in certain areas as the climate gets warmer, while others do not expect the number to change at all.[36,37] Any decrease in cold-related deaths, however, will not be enough to outweigh the increase in heat-related deaths.[38,39]

About the Indicator

This indicator shows the annual rate for deaths classified by medical professionals as "heat-related" each year in the United States, based on death certificate records. Every death is recorded on a death certificate, where a medical professional identifies the main cause of death (also known as the underlying cause), along with other conditions that contributed to the death. These causes are classified using a set of standard codes. Multiplying the annual number of deaths per U.S. population that year by one million will result in the death rate shown in Figure 1.

This indicator shows heat-related deaths using two methodologies. One method shows deaths for which excessive natural heat was stated as the underlying cause of death from 1979 to 2009. The other data series shows deaths for which heat was listed as either the underlying cause or a contributing cause, based on a broader set of data that at present can only be evaluated

(Continued on page 73)

Figure 1. Deaths Classified as "Heat-Related" in the United States, 1979–2009

This figure shows the annual rates for deaths classified as "heat-related" by medical professionals in the 50 states and the District of Columbia. The orange line shows deaths for which heat was listed as the main (underlying) cause.* The blue line shows deaths for which heat was listed as either the underlying or contributing cause of death during the months from May to September, based on a broader set of data that became available in 1999.

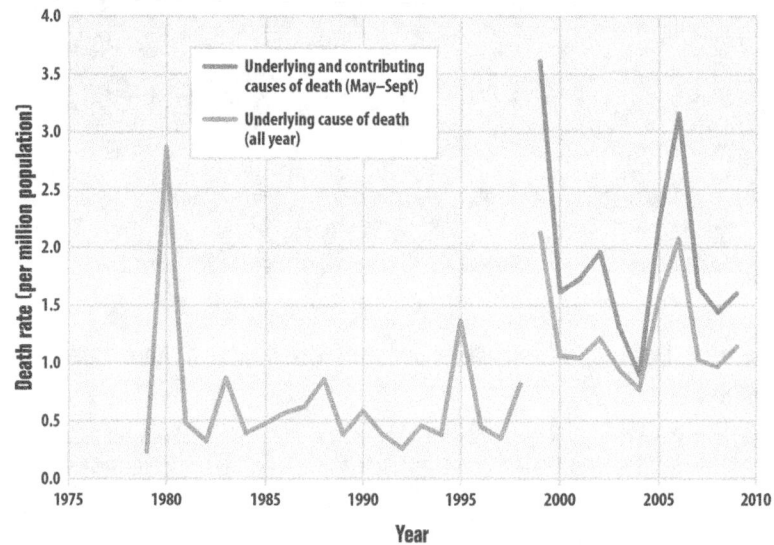

* Between 1998 and 1999, the World Health Organization revised the international codes used to classify causes of death. As a result, data from earlier than 1999 cannot easily be compared with data from 1999 and later.

Data source: CDC, 2012[40,41]

Key Points

- During the 31 years of data collection (1979–2009), the death rate as a direct result of exposure to heat (underlying cause of death) generally hovered around 0.5 deaths per million population, with spikes in certain years (see Figure 1). Overall, a total of over 7,000 Americans suffered heat-related deaths since 1979.

- For years where the two records overlap (1999–2009), accounting for those additional deaths in which heat was listed as a contributing factor results in a higher death rate—nearly double for some years—compared with the estimate that only includes deaths where heat was listed as the underlying cause. However, even this expanded metric does not necessarily capture the full extent of heat-related deaths.

- The indicator shows a peak in heat-related deaths in 2006, a year that was associated with widespread heat waves and was the second-hottest year on record in the contiguous 48 states (see the U.S. and Global Temperature indicator on p. 24).

- Considerable year-to-year variability in the data and certain limitations of this indicator make it difficult to determine whether the United States has experienced a meaningful increase or decrease in deaths classified as "heat-related" over time. Dramatic increases in heat-related deaths are closely associated with both the occurrence of hot temperatures and heat waves, though these deaths may not be reported as "heat-related" on death certificates. For example, studies of the 1995 heat wave event in Chicago (see example on p. 73) suggest that there were hundreds more deaths than were actually reported as "heat-related" on death certificates.

Example: Examining Heat-Related Deaths During the 1995 Chicago Heat Wave*

Many factors can influence the nature, extent, and timing of health consequences associated with extreme heat events.[42] Studies of heat waves are one way to better understand health impacts, but different methods can lead to very different estimates of heat-related deaths. For example, during a severe heat wave that hit Chicago between July 11 and July 27, 1995, 465 heat-related deaths were recorded on death certificates in Cook County.[43] However, studies that compared the total number of deaths during this heat wave (regardless of the recorded cause of death) with the long-term average of daily deaths found that the heat wave led to about 700 more deaths than would otherwise have been expected.[44] Differences in estimated heat-related deaths that result from different methods may be even larger when considering the entire nation and longer time periods.

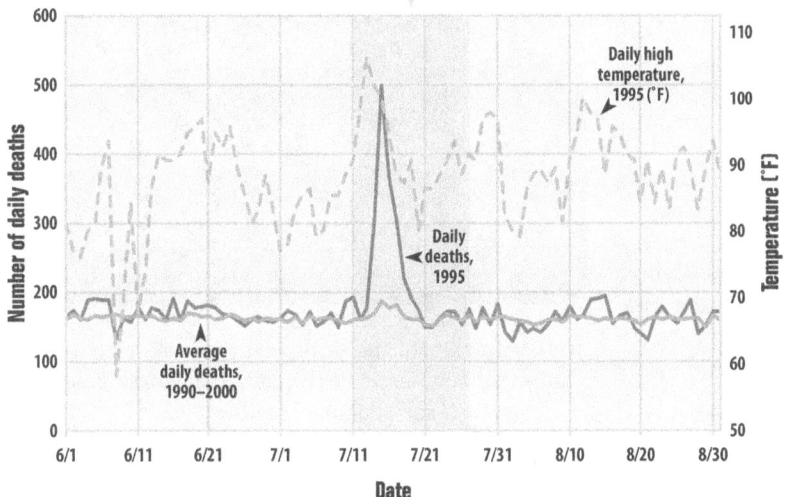

Cook County, July 11–27, 1995:
Excess deaths compared with this time period during an average year: about 700
Deaths classified as "heat-related" on death certificates (not shown here): 465

*This graph shows data for the Chicago Standard Metropolitan Statistical Area.

Data sources: CDC, 2012;[45] NOAA, 2012[46]

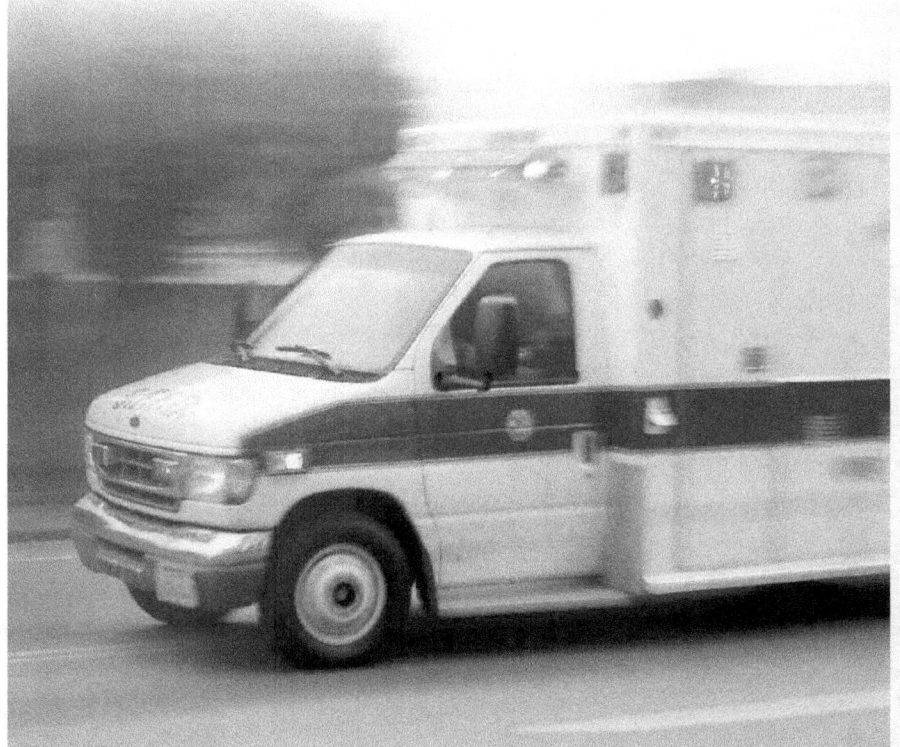

back to 1999. For example, in a case where cardiovascular disease was determined to be the underlying cause of death, heat could be listed as a contributing factor because it can make the individual more susceptible to the effects of this disease. Because excessive heat events are associated with summer months, the 1999–2009 analysis was limited to May through September.

Indicator Notes

Several factors influence the sensitivity of this indicator and its ability to estimate the true number of deaths associated with extreme heat events. It has been well-documented that many deaths associated with extreme heat are not identified as such by the medical examiner and might not be correctly coded on the death certificate. In many cases, the medical examiner might classify the cause of death as a cardiovascular or respiratory disease, not knowing for certain whether heat was a contributing factor, particularly if the death did not occur during a well-publicized heat wave. By studying how daily death rates vary with temperature in selected cities, scientists have found that extreme heat contributes to far more deaths than the official death certificates might suggest.[47] This is because the stress of a hot day can increase the chance of dying from a heart attack, other heart conditions, or respiratory diseases such as pneumonia.[48] These causes of death are much more common than heat-related illnesses such as heat stroke. Thus, this indicator very likely underestimates the number of deaths caused by exposure to heat.

Just because a death is classified as "heat-related" does not mean that high temperatures were the only factor that caused or contributed to the death. Pre-existing medical conditions can significantly increase an individual's vulnerability to heat. Other important factors, such as the overall vulnerability of the population, the extent to which people have adapted to higher temperatures, and the local climate and topography, can affect trends in "heat-related" deaths. Heat response measures such as early warning and surveillance systems, air conditioning, health care, public education, cooling centers during heat waves, infrastructure standards, and air quality management can also make a big difference in death rates. For example, after a 1995 heat wave, the city of Milwaukee developed a plan for responding to extreme heat conditions in the future; during the 1999 heat wave, heat-related deaths were roughly half of what would have been expected.[49]

Future development related to this indicator should focus on capturing *all* heat-related deaths, not just those with a reported link to heat stress, as well as examining heat-related illnesses more systematically.

Data Sources

Data for this indicator were provided by the U.S. Centers for Disease Control and Prevention (CDC). The 1979–2009 underlying cause data are publicly available through the CDC WONDER database at: http://wonder.cdc.gov/mortSQL.html. The 1999–2009 analysis was developed by CDC's Environmental Public Health Tracking Program, which provides a summary at: www.cdc.gov/nceh/tracking.

Climate Change Indicators and Human Health

This report provides several environmental and ecological indicators of observed change related to climate. Although climate change can affect human health in a number of direct and indirect ways, well-defined, consensus-based "health" indicators are limited.

Many societal and environmental factors can influence how climate change will affect health in any given community. These factors include the extent, frequency, and severity of climate change impacts; the ability of communities to prepare for and respond to the risks posed by climate change; and the vulnerability of the different populations living in the community.

Because the impacts of climate change on health are complex, often indirect, and dependent on multiple societal and environmental factors, the development of appropriate climate-related health indicators is challenging and still emerging. To ensure that response measures are effective and adverse health effects are avoided, it is important for climate-related health indicators to be clear, measurable, timely, and closely linked to changes in climate.[1,2]

Climate-related health indicators will be instrumental not only in tracking and measuring health impacts of climate change but also, more importantly, in identifying areas where the protection of public health is needed most. EPA plans to explore opportunities to

In 2011, The Centers for Disease Control and Prevention (CDC) introduced a suite of indicators to track the effects of climate change on human health through the Environmental Public Health Tracking Network (EPHTN). The network links standardized metrics from local, state, and national databases on environmental hazards and human health effects with climate information.[3] EPHTN is an emerging effort from which useful indicators may be leveraged to track potential direct and indirect health effects due to climate change. For more information about EPHTN, go to: www.cdc.gov/nceh/tracking.

work with climate and health experts to develop indicators that communicate the effects of climate change on health and society more broadly.

Key human health impacts and vulnerabilities associated with climate change include:

- A warmer climate will increase the risk of heat-related illness and death. A warmer climate is also expected to decrease the risk of cold-related illness and death.

- Climate change is expected to worsen conditions for air quality, including exposure to ground-level ozone, which can aggravate lung diseases and lead to premature death.

- Climate change will likely increase the frequency and strength of certain extreme events (such as floods, droughts, and storms) that threaten human safety and health.

- Changes in temperature and precipitation can spread or shift the geographic range of certain diseases and alter the seasons for pollen, affecting human exposure to infection, asthma, and other respiratory diseases.

- Vulnerable populations including the poor, the elderly, those already in poor health, the disabled, and indigenous populations are most at risk.

For more information about climate change impacts and human health, visit EPA's website: www.epa.gov/climatechange/impacts-adaptation/health.html.

Climate Change Resources

EPA's Climate Change website (www.epa.gov/climatechange) provides a good starting point for further exploration of this topic. From this site, you can:

- View the latest information about EPA's climate change indicators (www.epa.gov/climatechange/indicators) and download figures as well as accompanying technical documentation.

- Learn more about greenhouse gases and the science of climate change, discover the potential impacts of climate change on human health and ecosystems, read about how people can adapt to changes, and get up-to-date news.

- Read about greenhouse gas emissions, look through EPA's greenhouse gas inventories, and explore EPA's Greenhouse Gas Data Publication Tool.

- Learn about EPA's regulatory initiatives and partnership programs.

- Search EPA's database of frequently asked questions about climate change and ask your own questions. Explore a glossary of terms related to climate change, including many terms that appear in this report.

- Find out what you can do at home, on the road, at work, and at school to help reduce greenhouse gas emissions.

- Explore U.S. climate policy and climate economics.

- Find resources for educators and students.

Many other government and nongovernment websites also provide information about climate change. Here are some examples:

- The Intergovernmental Panel on Climate Change (IPCC) is the international authority on climate change science. The IPCC website (www.ipcc.ch/index.htm) summarizes the current state of scientific knowledge about climate change.

- The U.S. Global Change Research Program (www.globalchange.gov) is a multi-agency effort focused on improving our understanding of the science of climate change and its potential impacts on the United States through reports such as the National Climate Assessment.

- The National Academy of Sciences (http://nas-sites.org/americasclimatechoices) has developed many independent scientific reports on the causes of climate change, its impacts, and potential solutions. The National Academy's Koshland Science Museum (https://koshland-science-museum.org) provides an interactive online Earth Lab where people can learn more about these issues.

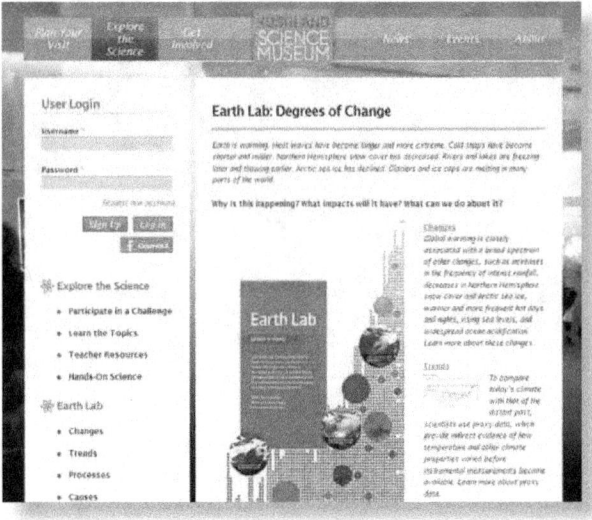

- The National Oceanic and Atmospheric Administration (NOAA) is charged with helping society understand, plan for, and respond to climate variability and change. Find out more about NOAA's climate indicators and other activities at: www.climate.gov.

- NOAA's National Climatic Data Center website (www.ncdc.noaa.gov/oa/ncdc.html) provides access to data that demonstrate the effects of climate change on weather, climate, and the oceans.

- The Centers for Disease Control and Prevention (CDC) provides extensive information about the relationship between climate change and public health at: www.cdc.gov/climateandhealth/default.htm.

- The U.S. Geological Survey's Climate and Land Use Change website (www.usgs.gov/climate_landuse) looks at the relationships between natural processes on the surface of the earth, ecological systems, and human activities.

- The National Aeronautics and Space Administration (NASA) maintains its own set of climate change indicators (http://climate.nasa.gov). Another NASA site (http://earthobservatory.nasa.gov/Features/EnergyBalance/page1.php) discusses the Earth's energy budget and how it relates to greenhouse gas emissions and climate change.

- The National Snow and Ice Data Center's website (http://nsidc.org/cryosphere) provides more information about ice and snow and how they influence and are influenced by climate change.

- The Woods Hole Oceanographic Institution's website (www.whoi.edu/main/climate-ocean) explains how climate change affects the oceans and how scientists measure these effects.

For more indicators of environmental condition, visit EPA's Report on the Environment (www.epa.gov/roe). This resource presents a wide range of indicators of national conditions and trends in air, water, land, human health, and ecological systems.

Endnotes

Introduction

1. IPCC (Intergovernmental Panel on Climate Change). 2007. Climate change 2007: Synthesis report (Fourth Assessment Report). Cambridge, United Kingdom: Cambridge University Press.

Summary of Key Points

1. IPCC (Intergovernmental Panel on Climate Change). 2007. Climate change 2007: Synthesis report (Fourth Assessment Report). Cambridge, United Kingdom: Cambridge University Press.

Greenhouse Gases

1. IPCC (Intergovernmental Panel on Climate Change). 1995. Climate change 1995: The science of climate change (Second Assessment Report). Cambridge, United Kingdom: Cambridge University Press.

2. IPCC (Intergovernmental Panel on Climate Change). 2007. Climate change 2007: The physical science basis (Fourth Assessment Report). Cambridge, United Kingdom: Cambridge University Press.

3. U.S. EPA (U.S. Environmental Protection Agency). 2012. Inventory of U.S. greenhouse gas emissions and sinks: 1990–2010. USEPA #EPA 430-R-12-001. www.epa.gov/climatechange/ghgemissions/usinventoryreport.html.

4. ibid.

5. ibid.

6. ibid.

7. ibid.

8. ibid.

9. World Resources Institute. 2012. Climate Analysis Indicators Tool (CAIT). Version 9.0. Accessed May 2012. http://cait.wri.org.

10. ibid.

11. ibid.

12. *EPICA Dome C, Antarctica: approximately 647,426 BC to 411,548 BC*
Siegenthaler, U., T. F. Stocker, E. Monnin, D. Lüthi, J. Schwander, B. Stauffer, D. Raynaud, J.M. Barnola, H. Fischer, V. Masson-Delmotte, and J. Jouzel. 2005. Stable carbon cycle-climate relationship during the late Pleistocene. Science 310(5752):1313–1317. Accessed May 15, 2007. ftp://ftp.ncdc.noaa.gov/pub/data/paleo/icecore/antarctica/epica_domec/edc-co2-650k-390k.txt.

Vostok Station, Antarctica: approximately 415,157 BC to 339 BC
Barnola, J.M., D. Raynaud, C. Lorius, and N.I. Barkov. 2003. Historical CO_2 record from the Vostok ice core. In: Trends: A compendium of data on global change. Oak Ridge, TN: U.S. Department of Energy. Accessed September 14, 2005. http://cdiac.ornl.gov/trends/co2/vostok.html.

EPICA Dome C, Antarctica: approximately 9002 BC to 1515 AD
Flückiger, J., E. Monnin, B. Stauffer, J. Schwander, T.F. Stocker, J. Chappellaz, D. Raynaud, and J.M. Barnola. 2002. High resolution Holocene N_2O ice core record and its relationship with CH_4 and CO_2. Global Biogeochem. Cycles 16(1):10–11. Accessed April 30, 2007. ftp://ftp.ncdc.noaa.gov/pub/data/paleo/icecore/antarctica/epica_domec/readme_flueckiger2002.txt.

Law Dome, Antarctica, 75-year smoothed: approximately 1010 AD to 1975 AD
Etheridge, D.M., L.P. Steele, R.L. Langenfelds, R.J. Francey, J.M. Barnola, and V.I. Morgan. 1998. Historical CO_2 records from the Law Dome DE08, DE08-2, and DSS ice cores. In: Trends: A compendium of data on global change. Oak Ridge, TN: U.S. Department of Energy. Accessed September 14, 2005. http://cdiac.ornl.gov/trends/co2/lawdome.html.

Siple Station, Antarctica: approximately 1744 AD to 1953 AD
Neftel, A., H. Friedli, E. Moor, H. Lötscher, H. Oeschger, U. Siegenthaler, and B. Stauffer. 1994. Historical CO_2 record from the Siple Station ice core. In: Trends: A compendium of data on global change. Oak Ridge, TN: U.S. Department of Energy. Accessed September 14, 2005. http://cdiac.ornl.gov/trends/co2/siple.html.

Mauna Loa, Hawaii: 1959 AD to 2011 AD
NOAA (National Oceanic and Atmospheric Administration). 2012. Annual mean CO_2 concentrations for Mauna Loa, Hawaii. Accessed May 10, 2012. ftp://ftp.cmdl.noaa.gov/ccg/co2/trends/co2_annmean_mlo.txt.

Barrow, Alaska: 1974 AD to 2011 AD
Cape Matatula, American Samoa: 1976 AD to 2011 AD
South Pole, Antarctica: 1976 AD to 2011 AD
NOAA (National Oceanic and Atmospheric Administration). 2012. Monthly mean CO_2 concentrations for Barrow, Alaska; Cape Matatula, American Samoa; and the South Pole. Accessed May 10, 2012. ftp://ftp.cmdl.noaa.gov/ccg/co2/in-situ.

Cape Grim, Australia: 1992 AD to 2006 AD
Shetland Islands, Scotland: 1993 AD to 2002 AD
Steele, L.P., P.B. Krummel, and R.L. Langenfelds. 2007. Atmospheric CO_2 concentrations (ppmv) derived from flask air samples collected at Cape Grim, Australia, and Shetland Islands, Scotland. Commonwealth Scientific and Industrial Research Organisation. Accessed January 20, 2009. http://cdiac.esd.ornl.gov/ftp/trends/co2/csiro.

Lampedusa Island, Italy: 1993 AD to 2000 AD
Chamard, P., L. Ciattaglia, A. di Sarra, and F. Monteleone. 2001. Atmospheric CO_2 record from flask measurements at Lampedusa Island. In: Trends: A compendium of data on global change. Oak Ridge, TN: U.S. Department of Energy. Accessed September 14, 2005. http://cdiac.ornl.gov/trends/co2/lampis.html.

13. *EPICA Dome C, Antarctica: approximately 646,729 BC to 1888 AD*
Spahni, R., J. Chappellaz, T.F. Stocker, L. Loulergue, G. Hausammann, K. Kawamura, J. Flückiger, J. Schwander, D. Raynaud, V. Masson-Delmotte, and J. Jouzel. 2005. Atmospheric methane and nitrous oxide of the late Pleistocene from Antarctic ice cores. Science 310(5752):1317–1321. Accessed May 15, 2007. ftp://ftp.ncdc.noaa.gov/pub/data/paleo/icecore/antarctica/epica_domec/edc-ch4-2005-650k.txt.

Vostok Station, Antarctica: approximately 415,172 BC to 346 BC
Petit, J.R., J. Jouzel, D. Raynaud, N.I. Barkov, J.M. Barnola, I. Basile, M. Bender, J. Chappellaz, M. Davis, G. Delaygue, M. Delmotte, V.M. Kotlyakov, M. Legrand, V. Lipenkov, C. Lorius, L. Pépin, C. Ritz, E. Saltzman, and M. Stievenard. 1999. Climate and atmospheric history of the past 420,000 years from the Vostok ice core, Antarctica. Nature 399:429–436. Accessed April 24, 2007. ftp://ftp.ncdc.noaa.gov/pub/data/paleo/icecore/antarctica/vostok/ch4nat.txt.

Greenland GISP2 ice core: approximately 87,798 BC to 8187 BC
Byrd Station, Antarctica: approximately 85,929 BC to 6748 BC
Greenland GRIP ice core: approximately 46,933 BC to 8129 BC
Blunier, T., and E.J. Brook. 2001. Timing of millennial-scale climate change in Antarctica and Greenland during the last glacial period. Science 291:109–112. Accessed September 13, 2005. ftp://ftp.ncdc.noaa.gov/pub/data/paleo/icecore/greenland/summit/grip/synchronization/readme_blunier2001.txt.

EPICA Dome C, Antarctica: approximately 8945 BC to 1760 AD
Flückiger, J., E. Monnin, B. Stauffer, J. Schwander, T.F. Stocker, J. Chappellaz, D. Raynaud, and J.M. Barnola. 2002. High resolution Holocene N_2O ice core record and its relationship with CH_4 and CO_2. Global Biogeochem. Cycles 16(1):10–11. Accessed April 24, 2007. ftp://ftp.ncdc.noaa.gov/pub/data/paleo/icecore/antarctica/epica_domec/readme_flueckiger2002.txt.

Law Dome, Antarctica: approximately 1008 AD to 1980 AD
Various Greenland locations: approximately 1075 AD to 1885 AD
Etheridge, D.M., L.P. Steele, R.J. Francey, and R.L. Langenfelds. 2002. Historical CH_4 records since about 1000 AD from ice core data. In: Trends: A compendium of data on global change. Oak Ridge, TN: U.S. Department of Energy. Accessed September 13, 2005. http://cdiac.ornl.gov/trends/atm_meth/lawdome_meth.html.

Greenland Site J: approximately 1598 AD to 1951 AD
WDCGG (World Data Centre for Greenhouse Gases). 2005. Atmospheric CH_4 concentrations for Greenland Site J. Accessed September 1, 2005. http://ds.data.jma.go.jp/gmd/wdcgg.

Cape Grim, Australia: 1984 AD to 2010 AD
NOAA (National Oceanic and Atmospheric Administration). 2011. Monthly mean CH_4 concentrations for Cape Grim, Australia. Accessed October 27, 2011. ftp://ftp.cmdl.noaa.gov/ccg/ch4/flask/month/ch4_cgo_surface-flask_1_ccgg_month.txt.

Mauna Loa, Hawaii: 1987 AD to 2011 AD
NOAA (National Oceanic and Atmospheric Administration). 2011. Monthly mean CH_4 concentrations for Mauna Loa, Hawaii. Accessed May 10, 2012. ftp://ftp.cmdl.noaa.gov/ccg/ch4/in-situ/mlo/ch4_mlo_surface-insitu_1_ccgg_month.txt.

Shetland Islands, Scotland: 1993 AD to 2001 AD
Steele, L.P., P.B. Krummel, and R.L. Langenfelds. 2002. Atmospheric CH_4 concentrations from sites in the CSIRO Atmospheric Research GASLAB air sampling network (October 2002 version). In: Trends: A compendium of data on global change. Oak Ridge, TN: U.S. Department of Energy. Accessed September 13, 2005. http://cdiac.esd.ornl.gov/trends/atm_meth/csiro/csiro-shetlandch4.html.

14. IPCC (Intergovernmental Panel on Climate Change). 2007. Climate change 2007: The physical science basis (Fourth Assessment Report). Cambridge, United Kingdom: Cambridge University Press.

15. IPCC (Intergovernmental Panel on Climate Change). 2007. Climate change 2007: Synthesis report (Fourth Assessment Report). Cambridge, United Kingdom: Cambridge University Press.

16. ibid.

17. AGAGE (Advanced Global Atmospheric Gases Experiment). 2011. ALE/GAGE/AGAGE data base. Accessed November 2011. http://agage.eas.gatech.edu/data.htm.

18. NOAA (National Oceanic and Atmospheric Administration). 2011. Halocarbons and other atmospheric trace species. Accessed October 2011. www.esrl.noaa.gov/gmd/hats.

19. Arnold, T., J. Mühle, P.K. Salameh, C.M. Harth, D.J. Ivy, and R.F. Weiss. 2012. Automated measurement of nitrogen trifluoride in ambient air. Analytical Chemistry 84(11):4798–4804.

20. Weiss, R.F., J. Mühle, P.K. Salameh, and C.M. Harth. 2008. Nitrogen trifluoride in the global atmosphere. Geophys. Res. Lett. 35:L20821.

21. IPCC (Intergovernmental Panel on Climate Change). 2007. Climate change 2007: Synthesis report (Fourth Assessment Report). Cambridge, United Kingdom: Cambridge University Press.

22. *Greenland GISP2 ice core: approximately 104,301 BC to 1871 AD*
Taylor Dome, Antarctica: approximately 30,697 BC to 497 BC
Sowers, T., R.B. Alley, and J. Jubenville. 2003. Ice core records of atmospheric N₂O covering the last 106,000 years. Science 301(5635):945–948. Accessed September 14, 2005. www.ncdc.noaa.gov/pub/data/paleo/icecore/antarctica/taylor/taylor_n2o.txt.

EPICA Dome C, Antarctica: approximately 9000 BC to 1780 AD
Flückiger, J., E. Monnin, B. Stauffer, J. Schwander, T.F. Stocker, J. Chappellaz, D. Raynaud, and J.M. Barnola. 2002. High resolution Holocene N₂O ice core record and its relationship with CH₄ and CO₂. Global Biogeochem. Cycles 16(1):10–11. Accessed September 14, 2005. ftp://ftp.ncdc.noaa.gov/pub/data/paleo/icecore/antarctica/epica_domec/readme_flueckiger2002.txt.

Antarctica: approximately 1756 AD to 1964 AD
Machida, T., T. Nakazawa, Y. Fujii, S. Aoki, and O. Watanabe. 1995. Increase in the atmospheric nitrous oxide concentration during the last 250 years. Geophys. Res. Lett. 22(21):2921–2924. Accessed September 8, 2005. ftp://daac.ornl.gov/data/global_climate/global_N_cycle/data/global_N_perturbations.txt.

Antarctica: approximately 1903 AD to 1976 AD
Battle, M., M. Bender, T. Sowers, P. Tans, J. Butler, J. Elkins, J. Ellis, T. Conway, N. Zhang, P. Lang, and A. Clarke. 1996. Atmospheric gas concentrations over the past century measured in air from firn at the South Pole. Nature 383:231–235. Accessed September 8, 2005. ftp://daac.ornl.gov/data/global_climate/global_N_cycle/data/global_N_perturbations.txt.

Cape Grim, Australia: 1979 AD to 2010 AD
AGAGE (Advanced Global Atmospheric Gases Experiment). 2011. Monthly mean N₂O concentrations for Cape Grim, Australia. Accessed May 10, 2012. http://ds.data.jma.go.jp/gmd/wdcgg.

South Pole, Antarctica: 1998 AD to 2011 AD
Barrow, Alaska: 1999 AD to 2011 AD
Mauna Loa, Hawaii: 2000 AD to 2011 AD
NOAA (National Oceanic and Atmospheric Administration). 2012. Monthly mean N₂O concentrations for Barrow, Alaska; Mauna Loa, Hawaii; and the South Pole. Accessed May 10, 2012. www.esrl.noaa.gov/gmd/hats/insitu/cats/cats_conc.html.

23. AGAGE (Advanced Global Atmospheric Gases Experiment). 2011. ALE/GAGE/AGAGE data base. Accessed November 2011. http://agage.eas.gatech.edu/data.htm.

24. Arnold, T., J. Mühle, P.K. Salameh, C.M. Harth, D.J. Ivy, and R.F. Weiss. 2012. Automated measurement of nitrogen trifluoride in ambient air. Analytical Chemistry 84(11):4798–4804.

25. NOAA (National Oceanic and Atmospheric Administration). 2011. Halocarbons and other atmospheric trace species. Accessed October 2011. www.esrl.noaa.gov/gmd/hats.

26. Weiss, R.F., J. Mühle, P.K. Salameh, and C.M. Harth. 2008. Nitrogen trifluoride in the global atmosphere. Geophys. Res. Lett. 35:L20821.

27. NOAA (National Oceanic and Atmospheric Administration). 2012. The NOAA Annual Greenhouse Gas Index. Accessed October 2012. www.esrl.noaa.gov/gmd/aggi.

Weather and Climate

1. NOAA (National Oceanic and Atmospheric Administration). 2009. Layers of the atmosphere. www.srh.noaa.gov/jetstream/atmos/layers.htm.

2. NOAA (National Oceanic and Atmospheric Administration). 2012. National Climatic Data Center. Accessed April 2012. www.ncdc.noaa.gov/oa/ncdc.html.

3. ibid.

4. ibid.

5. Karl, T.R., J.M. Melillo, and T.C. Peterson (eds.). 2009. Global climate change impacts in the United States. New York, NY: Cambridge University Press.

6. National Research Council. 2011. Climate stabilization targets: Emissions, concentrations, and impacts over decades to millennia. Washington, DC: National Academies Press.

7. Meehl, G.A., T.F. Stocker, W.D. Collins, P. Friedlingstein, A.T. Gaye, J.M. Gregory, A. Kitoh, R. Knutti, J.M. Murphy, A. Noda, S.C.B. Raper, I.G. Watterson, A.J. Weaver, and Z.-C. Zhao. 2007. Global climate projections. In: Climate change 2007: The physical science basis (Fourth Assessment Report). Cambridge, United Kingdom: Cambridge University Press.

8. Kunkel, K. 2012. Updated version of a figure that originally appeared in U.S. Climate Change Science Program's 2008 report: Synthesis and Assessment Product 3.3: Weather and climate extremes in a changing climate. www.climatescience.gov/Library/sap/sap3-3/final-report/sap3-3-final-Chapter2.pdf.

9. NOAA (National Oceanic and Atmospheric Administration). 2012. U.S. Climate Extremes Index. Accessed October 2012. www.ncdc.noaa.gov/oa/climate/research/cei/cei.html.

10. NOAA (National Oceanic and Atmospheric Administration). 2012. U.S. Climate Extremes Index. Accessed March 2012. www.ncdc.noaa.gov/oa/climate/research/cei/cei.html.

11. Meehl, G. A., C. Tebaldi, G. Walton, D. Easterling, and L. McDaniel. 2009. Relative increase of record high maximum temperatures compared to record low minimum temperatures in the U.S. Geophys. Res. Lett. 36:L23701.

12. CCSP (U.S. Climate Change Science Program). 2008. Synthesis and Assessment Product 3.3: Weather and climate extremes in a changing climate. www.climatescience.gov/Library/sap/sap3-3/final-report/sap3-3-final-Chapter2.pdf.

13. Meehl, G. A., C. Tebaldi, G. Walton, D. Easterling, and L. McDaniel. 2009. Relative increase of record high maximum temperatures compared to record low minimum temperatures in the U.S. Geophys. Res. Lett. 36:L23701.

14. NOAA (National Oceanic and Atmospheric Administration). 2012. National Climatic Data Center. Accessed April 2012. www.ncdc.noaa.gov/oa/ncdc.html.

15. ibid.

16. ibid.

17. Tebaldi, C., K. Hayhoe, J.M. Arblaster, and G.A. Meehl. 2006. Going to the extremes: An intercomparison of model-simulated historical and future changes in extreme events. Climatic Change 79:185–211.

18. NOAA (National Oceanic and Atmospheric Administration). 2012. U.S. Climate Extremes Index. Accessed March 2012. www.ncdc.noaa.gov/oa/climate/research/cei/cei.html.

19. NOAA (National Oceanic and Atmospheric Administration). 2012. Standardized Precipitation Index data files. Accessed March 2012. ftp://ftp.ncdc.noaa.gov/pub/data/cirs.

20. NOAA (National Oceanic and Atmospheric Administration). 2012. National Climatic Data Center. Personal communication: Analysis by Derek Arndt, April 2012.

21. CCSP (U.S. Climate Change Science Program). 2008. Synthesis and Assessment Product 3.3: Weather and climate extremes in a changing climate. www.climatescience.gov/Library/sap/sap3-3/final-report/sap3-3-final-Chapter2.pdf.

22. IPCC (Intergovernmental Panel on Climate Change). 2012. Managing the risks of extreme events and disasters to advance climate change adaptation. Cambridge, United Kingdom: Cambridge University Press. http://ipcc-wg2.gov/SREX.

23. Heim, R.R. 2002. A review of twentieth-century drought indices used in the United States. Bull. Amer. Meteor. Soc. 83(8):1149–1165.

24. NOAA (National Oceanic and Atmospheric Administration). 2012. State of the climate: Drought: September 2012. Accessed November 2012. www.ncdc.noaa.gov/sotc/drought/2012/9.

25. NOAA (National Oceanic and Atmospheric Administration). 2012. National Climatic Data Center. Accessed January 2012. www.ncdc.noaa.gov/oa/ncdc.html.

26. National Drought Mitigation Center. 2012. Drought Monitor archives. Accessed January 2012. http://droughtmonitor.unl.edu/archive.html.

27. CCSP (U.S. Climate Change Science Program). 2008. Synthesis and Assessment Product 3.3: Weather and climate extremes in a changing climate. www.climatescience.gov/Library/sap/sap3-3/final-report/sap3-3-final-Chapter2.pdf.

28. ibid.

29. IPCC (Intergovernmental Panel on Climate Change). 2012. Managing the risks of extreme events and disasters to advance climate change adaptation. Cambridge, United Kingdom: Cambridge University Press. http://ipcc-wg2.gov/SREX.

30. Knutson, T.R. 2012 update to data originally published in: Knutson, T.R., J.L. McBride, J. Chan, K. Emanuel, G. Holland, C. Landsea, I. Held, J.P. Kossin, A.K. Srivastava, and M. Sugi. 2010. Tropical cyclones and climate change. Nature Geosci. 3:157–163.

31. NOAA (National Oceanic and Atmospheric Administration). 2012. The 2011 North Atlantic hurricane season: A climate perspective. www.cpc.ncep.noaa.gov/products/expert_assessment/hurrsummary_2011.pdf.

32. IPCC (Intergovernmental Panel on Climate Change). 2012. Managing the risks of extreme events and disasters to advance climate change adaptation. Cambridge, United Kingdom: Cambridge University Press. http://ipcc-wg2.gov/SREX.

33. Emanuel, K.A. 2012 update to data originally published in: Emanuel, K.A. 2007. Environmental factors affecting tropical cyclone power dissipation. J. Climate 20(22):5497–5509.

34. Knutson, T.R., J.L. McBride, J. Chan, K. Emanuel, G. Holland, C. Landsea, I. Held, J.P. Kossin, A.K. Srivastava, and M. Sugi. 2010. Tropical cyclones and climate change. Nature Geosci. 3:157–163.

Oceans

1. Khatiwala, S., F. Primeau, and T. Hall. 2009. Reconstruction of the history of anthropogenic CO_2 concentrations in the ocean. Nature 462:346–349.

2. Levitus, S., J. Antonov, and T. Boyer. 2005. Warming of the world ocean, 1955–2003. Geophys. Res. Lett. 32:L02604.

3. ibid.

4. CSIRO (Commonwealth Scientific and Industrial Research Organisation). 2012. Data downloads: Sea level and ocean heat content. Accessed February 2012. www.cmar.csiro.au/sealevel/sl_data_cmar.html.

5. JAMSTEC (Japan Agency for Marine-Earth Science and Technology). 2012 update to data originally published in: Ishii, M., and M. Kimoto. 2009. Reevaluation of historical ocean heat content variations with time-varying XBT and MBT depth bias corrections. J. Oceanogr. 65:287–299.

6. NOAA (National Oceanic and Atmospheric Administration). 2012. Global ocean heat and salt content. Accessed February 2012. www.nodc.noaa.gov/OC5/3M_HEAT_CONTENT.

7. For example, see Ostrander, G.K., K.M. Armstrong, E.T. Knobbe, D. Gerace, and E.P. Scully. 2000. Rapid transition in the structure of a coral reef community: The effects of coral bleaching and physical disturbance. Proc. Natl. Acad. Sci. 97(10):5297–5302.

8. Pratchett, M.S., S.K. Wilson, M.L. Berumen, and M.I. McCormick. 2004. Sublethal effects of coral bleaching on an obligate coral feeding butterflyfish. Coral Reefs 23(3):352–356.

9. Trenberth, K.E., P.D. Jones, P. Ambenje, R. Bojariu, D. Easterling, A. Klein Tank, D. Parker, F. Rahimzadeh, J.A. Renwick, M. Rusticucci, B. Soden, and P. Zhai. 2007. Observations: Surface and atmospheric climate change. In: Climate change 2007: The physical science basis (Fourth Assessment Report). Cambridge, United Kingdom: Cambridge University Press.

10. NOAA (National Oceanic and Atmospheric Administration). 2012. Extended reconstructed sea surface temperature (ERSST.v3b). National Climatic Data Center. Accessed April 2012. www.ncdc.noaa.gov/ersst.

11. UK Met Office. 2012. Hadley Centre, HadISST 1.1: Global sea ice coverage and sea surface temperature (1870–present). NCAS British Atmospheric Data Centre. Accessed May 2012. http://badc.nerc.ac.uk/view/badc.nerc.ac.uk__ATOM__dataent_hadisst.

12. CSIRO (Commonwealth Scientific and Industrial Research Organisation). 2012 update to data originally published in: Church, J.A., and N.J. White. 2011. Sea-level rise from the late 19th to the early 21st century. Surv. Geophys. 32:585–602.

13. NOAA (National Oceanic and Atmospheric Administration). 2012. Laboratory for Satellite Altimetry: Sea level rise. Accessed May 2012. http://ibis.grdl.noaa.gov/SAT/SeaLevelRise/LSA_SLR_timeseries_global.php.

14. Titus, J.G., E.K. Anderson, D.R. Cahoon, S. Gill, R.E. Thieler, and J.S. Williams. 2009. Coastal sensitivity to sea-level rise: A focus on the Mid-Atlantic region. U.S. Climate Change Science Program and the Subcommittee on Global Change Research. www.climatescience.gov/Library/sap/sap4-1/final-report/default.htm.

15. University of Colorado at Boulder. 2012. Sea level change: 2012 release #2. Accessed May 2012. http://sealevel.colorado.edu.

16. NOAA (National Oceanic and Atmospheric Administration). 2012 update to data originally published in: NOAA. 2001. Sea level variations of the United States 1854–1999. NOAA Technical Report NOS CO-OPS 36. http://tidesandcurrents.noaa.gov/publications/techrpt36.pdf.

17. IPCC (Intergovernmental Panel on Climate Change). 2007. Climate change 2007: The physical science basis (Fourth Assessment Report). Cambridge, United Kingdom: Cambridge University Press.

18. Wootton, J.T., C.A. Pfister, and J.D. Forester. 2008. Dynamic patterns and ecological impacts of declining ocean pH in a high-resolution multi-year dataset. Proc. Natl. Acad. Sci. 105(48):18848–18853.

19. Feely, R.A., S.C. Doney, and S.R. Cooley. 2009. Ocean acidification: Present conditions and future changes in a high-CO_2 world. Oceanography 22(4):36–47.

20. Bates, N.R., M.H.P. Best, K. Neely, R. Garley, A.G. Dickson, and R.J. Johnson. 2012. Detecting anthropogenic carbon dioxide uptake and ocean acidification in the North Atlantic Ocean. Biogeosciences Discuss. 9:989–1019.

21. González-Dávila, M. 2012 update to data originally published in: González-Dávila, M., J.M. Santana-Casiano, M.J. Rueda, and O. Llinás. 2010. The water column distribution of carbonate system variables at the ESTOC site from 1995 to 2004. Biogeosciences Discuss. 7:1995–2032.

22. University of Hawaii. 2012. Hawaii Ocean Time-Series. Accessed June 2012. http://hahana.soest.hawaii.edu/hot/products/HOT_surface_CO2.txt.

23. Recreated from Environment Canada. 2008. The pH scale. www.ec.gc.ca/eau-water/default.asp?lang=En&n=FDF30C16-1.

24. Feely, R.A., S.C. Doney, and S.R. Cooley. 2009. Ocean acidification: Present conditions and future changes in a high-CO_2 world. Oceanography 22(4):36–47.

25. ibid.

26. ibid.

27. Bates, N.R., M.H.P. Best, K. Neely, R. Garley, A.G. Dickson, and R.J. Johnson. 2012. Detecting anthropogenic carbon dioxide uptake and ocean acidification in the North Atlantic Ocean. Biogeosciences Discuss. 9:989–1019.

28. González-Dávila, M., J.M. Santana-Casiano, M.J. Rueda, and O. Llinás. 2010. The water column distribution of carbonate system variables at the ESTOC site from 1995 to 2004. Biogeosciences Discuss. 7:1995–2032.

29. Dore, J.E., R. Lukas, D.W. Sadler, M.J. Church, and D.M. Karl. 2009. Physical and biogeochemical modulation of ocean acidification in the central North Pacific. Proc. Natl. Acad. Sci. USA 106:12235–12240.

Snow and Ice

1. UNEP (United Nations Environment Programme). 2007. Global outlook for ice and snow. Cartographer: Hugo Ahlenius, UNEP/GRID-Arendal. www.unep.org/geo/geo_ice.

   Map based on the following data sources:

   Armstrong, R.L., and M.J. Brodzik. 2005. Northern Hemisphere EASE-Grid weekly snow cover and sea ice extent version 3. National Snow and Ice Data Center.

   Armstrong, R.L., M.J. Brodzik, K. Knowles, and M. Savoie. 2005. Global monthly EASE-Grid snow water equivalent climatology. National Snow and Ice Data Center.

   Brown, J., O.J. Ferrians, Jr., J.A. Heginbottom, and E.S. Melnikov. 2001. Circum-Arctic map of permafrost and ground-ice conditions. National Snow and Ice Data Center/World Data Center for Glaciology.

   National Geospatial-Intelligence Agency. 2000. Vector map level 0. http://geoengine.nima.mil/ftpdir/archive/vpf_data/v0soa.tar.gz.

   Stroeve, J., and W. Meier. 2005. Sea ice trends and climatologies from SMMR and SSM/I. National Snow and Ice Data Center. http://nsidc.org/data/smmr_ssmi_ancillary/monthly_means.html.

2. NASA (National Aeronautics and Space Administration). 2012. Sea ice yearly minimum 1979–2011. NASA/Goddard Space Flight Center Scientific Visualization Studio. Supplemented by personal communication with NASA in November 2012 to obtain a draft image for 2012. http://svs.gsfc.nasa.gov/vis/a000000/a003800/a003893.

3. NSIDC (National Snow and Ice Data Center). 2012. Arctic sea ice news and analysis. http://nsidc.org/icelights/arctic-sea-ice.

4. Comiso, J. 2012. Large decadal decline of the Arctic multiyear ice cover. J. Climate 25(4):1176–1193.

5. NSIDC (National Snow and Ice Data Center). 2012. Archived monthly sea ice data and images. Accessed October 2012. http://nsidc.org/data/seaice_index/archives/index.html.

6. NSIDC (National Snow and Ice Data Center). 2012. Arctic sea ice news and analysis. October 2, 2012. http://nsidc.org/arcticseaicenews/2012/10/poles-apart-a-record-breaking-summer-and-winter.

7. National Research Council. 2011. Climate stabilization targets: Emissions, concentrations, and impacts over decades to millennia. Washington, DC: National Academies Press.

8. Post, A. 1958. McCall Glacier. Glacier photograph collection. Boulder, Colorado: National Snow and Ice Data Center/World Data Center for Glaciology. http://nsidc.org/data/g00472.html.

9. Nolan, M. 2003. McCall Glacier. Glacier photograph collection. Boulder, Colorado: National Snow and Ice Data Center/World Data Center for Glaciology. http://nsidc.org/data/g00472.html.

10. Lemke, P., J. Ren, R.B. Alley, I. Allison, J. Carrasco, G. Flato, Y. Fujii, G. Kaser, P. Mote, R.H. Thomas, and T. Zhang. 2007. Observations: Changes in snow, ice and frozen ground. In: Climate change 2007: The physical science basis (Fourth Assessment Report). Cambridge, United Kingdom: Cambridge University Press.

11. WGMS (World Glacier Monitoring Service). 2011. Glacier mass balance bulletin no. 11 (2008–2009). Zemp, M., S.U. Nussbaumer, I. Gärtner-Roer, M. Hoelzle, F. Paul, and W. Haeberli (eds.). ICSU(WDS)/IUGG(IACS)/UNEP/UNESCO/WMO. Zurich, Switzerland: World Glacier Monitoring Service. www.wgms.ch/mbb/mbb11/wgms_2011_gmbb11.pdf.

12. WGMS (World Glacier Monitoring Service). 2012. Preliminary glacier mass balance data 2009/2010. www.wgms.ch/mbb/sum10.html.

13. USGS (U.S. Geological Survey). 2012. Water resources of Alaska—glacier and snow program, benchmark glaciers. Accessed June 2012. http://ak.water.usgs.gov/glaciology.

14. GCRP (U.S. Global Change Research Program). 2009. Global climate change impacts in the United States. www.globalchange.gov/what-we-do/assessment/previous-assessments/global-climate-change-impacts-in-the-us-2009.

15. Magnuson, J.J., D.M. Robertson, B.J. Benson, R.H. Wynne, D.M. Livingstone, T. Arai, R.A. Assel, R.G. Barry, V. Card, E. Kuusisto, N.G. Granin, T.D. Prowse, K.M. Stewart, and V.S. Vuglinski. 2000. Historical trends in lake and river ice cover in the Northern Hemisphere. Science 289:1743–1746. Errata 2001. Science 291:254.

16. NSIDC (National Snow and Ice Data Center). 2011. Global lake and river ice phenology. Internal development version accessed by NSIDC staff, December 2011. http://nsidc.org/data/lake_river_ice.

17. ibid.

18. ibid.

19. Kunkel, K.E., M. Palecki, L. Ensor, K.G. Hubbard, D. Robinson, K. Redmond, and D. Easterling. 2009. Trends in twentieth-century U.S. snowfall using a quality-controlled dataset. Journal of Atmospheric and Oceanic Technology 26:33–44.

20. Feng, S. 2012 update to data originally published in: Feng, S., and Q. Hu. 2007. Changes in winter snowfall/precipitation ratio in the contiguous United States. Journal of Geophysical Research 112:D15109.

21. Kunkel, K.E., M. Palecki, L. Ensor, K.G. Hubbard, D. Robinson, K. Redmond, and D. Easterling. 2009. Trends in twentieth-century U.S. snowfall using a quality-controlled dataset. Journal of Atmospheric and Oceanic Technology 26:33–44.

22. ibid.

23. Feng, S., and Q. Hu. 2007. Changes in winter snowfall/precipitation ratio in the contiguous United States. Journal of Geophysical Research 112:D15109.

24. Rutgers University Global Snow Lab. 2012. Area of extent data: North America (no Greenland). Accessed January 2012. http://climate.rutgers.edu/snowcover.

25. Rutgers University Global Snow Lab. 2012. Area of extent data: North America (no Greenland). Accessed February 2012. http://climate.rutgers.edu/snowcover.

26. Mote, P.W., A.F. Hamlet, M.P. Clark, and D.P. Lettenmaier. 2005. Declining mountain snowpack in Western North America. Bull. Amer. Meteor. Soc. 86(1):39–49.

27. Mote, P.W. 2009 update to data originally published in: Mote, P.W., A.F. Hamlet, M.P. Clark, and D.P. Lettenmaier. 2005. Declining mountain snowpack in Western North America. Bull. Amer. Meteor. Soc. 86(1):39–49.

28. Mote, P.W., A.F. Hamlet, M.P. Clark, and D.P. Lettenmaier. 2005. Declining mountain snowpack in Western North America. Bull. Amer. Meteor. Soc. 86(1):39–49.

Society and Ecosystems

1. USGS (U.S. Geological Survey). 2012. Analysis of data from the National Water Information System.

2. ibid.

3. ibid.

4. Falcone, J.A., D.M. Carlisle, D.M. Wolock, and M.R. Meador. 2010. GAGES: A stream gage database for evaluating natural and altered flow conditions in the conterminous United States. Ecology 91(2):621.

5. Arbes, S.J., Jr., P.J. Gergen, L. Elliott, and D.C. Zeldin. 2005. Prevalences of positive skin test responses to 10 common allergens in the U.S. population: Results from the third National Health and Nutrition Examination Survey. J. Allergy Clin. Immunol. 116(2):377–383.

6. Schappert, S.M., and E.A. Rechtsteiner. 2011. Ambulatory medical care utilization estimates for 2007. National Center for Health Statistics. Vital Health Stat 13(169). www.cdc.gov/nchs/data/series/sr_13/sr13_169.pdf.

7. Arbes, S.J., Jr., P.J. Gergen, L. Elliott, and D.C. Zeldin. 2005. Prevalences of positive skin test responses to 10 common allergens in the U.S. population: Results from the third National Health and Nutrition Examination Survey. J. Allergy Clin. Immunol. 116(2):377–383.

8. National Institute of Allergy and Infectious Diseases. 2011. Pollen allergy. www.niaid.nih.gov/topics/allergicDiseases/understanding/pollenallergy/Pages/default.aspx.

9. Wayne, P., S. Foster, J. Connolly, F. Bazzaz, and P. Epstein. 2002. Production of allergenic pollen by ragweed (Ambrosia artemisiifolia L.) is increased in CO_2-enriched atmospheres. Ann. Allergy. Asthma Im. 88:279–282.

10. Ziska, L., K. Knowlton, C. Rogers, D. Dalan, N. Tierney, M. Elder, W. Filley, J. Shropshire, L.B. Ford, C. Hedberg, P. Fleetwood, K.T. Hovanky, T. Kavanaugh, G. Fulford, R.F. Vrtis, J.A. Patz, J. Portnoy, F. Coates, L. Bielory, and D. Frenz. 2012 update to data originally published in: Ziska, L., K. Knowlton, C. Rogers, D. Dalan, N. Tierney, M. Elder, W. Filley, J. Shropshire, L.B. Ford, C. Hedberg, P. Fleetwood, K.T. Hovanky, T. Kavanaugh, G. Fulford, R.F. Vrtis, J.A. Patz, J. Portnoy, F. Coates, L. Bielory, and D. Frenz. 2011. Recent warming by latitude associated with increased length of ragweed pollen season in central North America. PNAS 108:4248–4251.

11. IPCC (Intergovernmental Panel on Climate Change). 2007. Climate change 2007: Synthesis report (Fourth Assessment Report). www.ipcc.ch/publications_and_data/ar4/syr/en/contents.html.

12. Ziska, L., K. Knowlton, C. Rogers, D. Dalan, N. Tierney, M. Elder, W. Filley, J. Shropshire, L.B. Ford, C. Hedberg, P. Fleetwood, K.T. Hovanky, T. Kavanaugh, G. Fulford, R.F. Vrtis, J.A. Patz, J. Portnoy, F. Coates, L. Bielory, and D. Frenz. 2011. Recent warming by latitude associated with increased length of ragweed pollen season in central North America. PNAS 108:4248–4251.

13. ibid.

14. Easterling, W.E., P.K. Aggarwal, P. Batima, K.M. Brander, L. Erda, S.M. Howden, A. Kirilenko, J. Morton, J.-F. Soussana, J. Schmidhuber, and F.N. Tubiello. 2007. Food, fibre and forest products. In: Climate change 2007: Impacts, adaptation and vulnerability (Fourth Assessment Report). Cambridge, United Kingdom: Cambridge University Press.

15. Kunkel, K.E. 2012 update to data originally published in: Kunkel, K.E., D.R. Easterling, K. Hubbard, and K. Redmond. 2004. Temporal variations in frost-free season in the United States: 1895–2000. Geophys. Res. Lett. 31:L03201.

16. ibid.

17. ibid.

18. ibid.

19. Rosenzweig, C., G. Casassa, D.J. Karoly, A. Imeson, C. Liu, A. Menzel, S. Rawlins, T.L. Root, B. Seguin, and P. Tryjanowski. 2007. Assessment of observed changes and responses in natural and managed systems. In: Climate change 2007: Impacts, adaptation, and vulnerability (Fourth Assessment Report). Cambridge, United Kingdom: Cambridge University Press.

20. Schwartz, M.D., R. Ahas, and A. Aasa. 2006. Onset of spring starting earlier across the Northern Hemisphere. Glob. Chang. Biol. 12:343–351.

21. Schwartz, M.D. 2011 update to data originally published in: Schwartz, M.D., R. Ahas, and A. Aasa. 2006. Onset of spring starting earlier across the Northern Hemisphere. Glob. Chang. Biol. 12:343–351.

22. ibid.

23. For example, see: Schwartz, M.D., R. Ahas, and A. Aasa. 2006. Onset of spring starting earlier across the Northern Hemisphere. Glob. Chang. Biol. 12:343–351.

24. McCabe, G.J., T.R. Ault, B.I. Cook, J.L. Betancourt, and M.D. Schwartz. 2011. Influences of the El Niño Southern Oscillation and the Pacific Decadal Oscillation on the timing of the North American spring. Int. J. Climatol. (online).

25. National Audubon Society. 2009. Northward shifts in the abundance of North American birds in early winter: A response to warmer winter temperatures? www.audubon.org/bird/bacc/techreport.html.

26. ibid.

27. ibid.

28. ibid.

29. ibid.

30. ibid.

31. ibid.

32. Hansen, J., M. Sato, and R. Ruedy. 2012. Perception of climate change. PNAS. Published online: August 6, 2012.

33. GCRP (U.S. Global Change Research Program). 2009. Global climate change impacts in the United States. www.globalchange.gov/usimpacts.

34. IPCC (Intergovernmental Panel on Climate Change). 2007. Climate change 2007: Impacts, adaptation, and vulnerability (Fourth Assessment Report). Cambridge, United Kingdom: Cambridge University Press.

35. Zanobetti, A., M.S. O'Neill, C.J. Gronlund, and J.D. Schwartz. 2012. Summer temperature variability and long-term survival among elderly people with chronic disease. PNAS 109(17):6608–6613.

36. IPCC (Intergovernmental Panel on Climate Change). 2007. Climate change 2007: Impacts, adaptation, and vulnerability (Fourth Assessment Report). Cambridge, United Kingdom: Cambridge University Press.

37. Medina-Ramón, M., and J. Schwartz. 2007. Temperature, temperature extremes, and mortality: A study of acclimatization and effect modification in 50 U.S. cities. Occup. Envi. Med. 64(12):827–833.

38. ibid.

39. IPCC (Intergovernmental Panel on Climate Change). 2007. Climate change 2007: Impacts, adaptation, and vulnerability (Fourth Assessment Report). Cambridge, United Kingdom: Cambridge University Press.

40. CDC (U.S. Centers for Disease Control and Prevention). 2012. CDC WONDER database. Accessed August 2012. http://wonder.cdc.gov/mortSQL.html.

41. CDC (U.S. Centers for Disease Control and Prevention). 2012. Indicator: Heat-related mortality. National Center for Health Statistics. Annual national totals provided by National Center for Environmental Health staff in August 2012. http://ephtracking.cdc.gov/showIndicatorPages.action.

42. Anderson, G.B., and M.L. Bell. 2011. Heat waves in the United States: Mortality risk during heat waves and effect modification by heat wave characteristics in 43 U.S. communities. Environ. Health Perspect. 119(2):210–218.

43. CDC (U.S. Centers for Disease Control and Prevention). 1995. Heat-related mortality – Chicago, July 1995. Morbidity and Mortality Weekly Report 44(31):577–579.

44. NRC (National Research Council). 2011. Climate stabilization targets: Emissions, concentrations, and impacts over decades to millennia. Washington, DC: National Academies Press.

45. CDC (U.S. Centers for Disease Control and Prevention). 2012. CDC WONDER database. Accessed August 2012. http://wonder.cdc.gov/mortSQL.html.

46. NOAA (National Oceanic and Atmospheric Administration). 2012. National Climatic Data Center. Accessed August 2012. www1.ncdc.noaa.gov/pub/orders/72774.dat.

47. Medina-Ramón, M., and J. Schwartz. 2007. Temperature, temperature extremes, and mortality: A study of acclimatization and effect modification in 50 U.S. cities. Occup. Envi. Med. 64(12):827–833.

48. Kaiser, R., A. Le Tertre, J. Schwartz, C.A. Gotway, W.R. Daley, and C.H. Rubin. 2007. The effect of the 1995 heat wave in Chicago on all-cause and cause-specific mortality. Am. J. Public Health 97(Supplement 1):S158–S162.

49. Weisskopf, M.G., H.A. Anderson, S. Foldy, L.P. Hanrahan, K. Blair, T.J. Torok, and P.D. Rumm. 2002. Heat wave morbidity and mortality, Milwaukee, Wis., 1999 vs. 1995: An improved response? Am. J. Public Health 92:830–833.

Climate Change Indicators and Human Health

1. English, P.B., A. H. Sinclair, Z. Ross, H. Anderson, V. Boothe, C. Davis, K. Ebi, B. Kagey, K. Malecki, R. Shultz, and E. Simms. 2009. Environmental health indicators of climate change for the United States: Findings from the State Environmental Health Indicator Collaborative. Environ. Health Perspect. 117(11):1673–1681.

2. Portier, C.J., T.K. Thigpen, S.R. Carter, C.H. Dilworth, A.E. Grambsch, J. Gohlke, J. Hess, S.N. Howard, G. Luber, J.T. Lutz, T. Maslak, N. Prudent, M. Radtke, J.P. Rosenthal, T. Rowles, P.A. Sandifer, J. Scheraga, P.J. Schramm, D. Strickman, J.M. Trtanj, and P.-Y. Whung. 2010. A human health perspective on climate change: A report outlining the research needs on the human health effects of climate change. Research Triangle Park, NC: Environmental Health Perspectives/National Institute of Environmental Health Sciences.

3. GCRP (U.S. Global Change Research Program). 2011. Societal indicators for the National Climate Assessment. NCA Report Series, Volume 5c. http://library.globalchange.gov/national-climate-assessment-societal-indicators-workshop-report.

www.ingramcontent.com/pod-product-compliance
Lightning Source LLC
Chambersburg PA
CBHW080644180526
45168CB00008B/3298